Genetic Testing and the Use of Information

Genetic Testing and the Use of Information

Clarisa Long, Editor

The AEI Press
Publisher for the American Enterprise Institute
WASHINGTON, D.C.
1999

Distributed to the Trade by National Book Network, 15200 NBN Way, Blue Ridge Summit, PA 17214. To order call toll free 1-800-462-6420 or 1-717-794-3800. For all other inquiries please contact the AEI Press, 1150 Seventeenth Street, N.W., Washington, D.C. 20036 or call 1-800-862-5801.

Library of Congress Cataloging-in-Publication Data

Genetic testing and the use of information / Clarisa Long, editor.
p. cm.
Includes index.
ISBN 0-8447-4110-8 (pbk.: alk. paper).—ISBN 0-8447-4109-4 (cloth: alk. paper)
1. Human chromosome abnormalities—Diagnosis—Moral and ethical aspects. I. Long, Clarisa.
RB155.6.G46 1999
616'.042—dc21 99-30858
CIP

1 3 5 7 9 10 8 6 4 2

ISBN 978-0-8447-4110-9

The AEI Press
Publisher for the American Enterprise Institute
1150 17th Street, N.W.
Washington, D.C. 20036

Contents

Contributors

Ellen Wright Clayton, M.D., J.D., is an associate professor of pediatrics and an associate professor of law in Vanderbilt University's School of Medicine and Law. She has conducted empirical research and written numerous articles on the legal, ethical, and social impact of new developments in genetics. Dr. Clayton's research interests include family law, child abuse and neglect, and factors that lead patients to sue their physicians. She is the president of the American Society of Law, Medicine, and Ethics and a member of the National Advisory Council for the National Human Genome Research Institute and the Social Issues Committee of the American Society of Human Genetics.

David Korn, M.D., is the senior vice president for biomedical and health sciences research at the Association of American Medical Colleges and the vice president, dean, and professor of pathology emeritus at the Stanford University School of Medicine. From 1984 to 1995, he was Carl and Elizabeth Naumann Professor and dean of the Stanford University School of Medicine; from 1986 to 1995 he was the vice president of Stanford University. Dr. Korn was a professor and the chairman of the department of pathology at Stanford University since 1968. He is a founder of the California Transplant Donor Network and was the chairman of the National Cancer Advisory Board, appointed by President Reagan. Dr. Korn is the author of numerous scientific articles and has been a member of the editorial boards of the *American Journal of Pathology*, the *Journal of Biological Chemistry*, and *Human Pathology*.

Clarisa Long, J.D., is the Abramson Fellow at the American Enterprise Institute and a research fellow at the Kennedy School of Government at Harvard University. She is the coauthor of *Genomic Information and Intellectual Property Rights: A Clash of Cultures* (1999) and of *U.S. Foreign Policy and Intellectual Property Protection in Latin America* (1997). She has been a molecular biologist at the Centre for Gene Technology in New Zealand and at the National Cancer Institute of the National Institutes of Health and is an intellectual property attorney in private practice.

Philip R. Reilly, M.D., J.D., is the executive director of the Shriver Center for Mental Retardation. He is also a director of the American Society of Human Genetics; a senior scientist of social science, ethics, and law at the Shriver Center; a clinical assistant in the department of neurology of Massachusetts General Hospital; and an adjunct professor of biology at Brandeis University. Dr. Reilly has held positions at the Harvard University School of Medicine, Powers & Hall, Vivigen, Inc., Boston City Hospital, the University of Houston Law Center, and the University of Texas Health Science Center. He is the author of *The Surgical Solution: A History of Involuntary Sterilization in the United States* (1991), *To Do No Harm* (1987), and *Genetics, Law, and Social Policy* (1977). Dr. Reilly is a frequent contributor to several scholarly journals, including the *Journal of the American Medical Association*, the *American Journal of Human Genetics*, and the *Journal of the National Cancer Institute.*

Karen Rothenberg, J.D., M.P.H., is the Marjorie Cook Professor of Law and the founding director of the law and health care program at the University of Maryland School of Law. She was a special assistant to the director of the Office of Research on Women's Health at the National Institutes of Health and practiced law with Covington and Burling. Ms. Rothenberg recently completed a series of studies on legislative approaches to genetic information in health insurance and the workplace as well as an extensive research project on the evolution from exclusion to inclusion of women in clinical research. She has contributed articles on AIDS, women's health, genetic testing, the right to forgo treatment, emergency care, and the new reproductive technologies to several scholarly journals, and she lectures extensively on legal issues in health care. Ms. Rothenberg is the coeditor-in-chief of the *Journal of Law, Medicine, and Ethics* and the coeditor of *Women and Prenatal Testing: Facing the Challenges of Genetic Technology.*

Michael S. Watson, Ph.D., is the vice president for laboratory affairs and a director of the American College of Medical Genetics. He is also the chairman of the Laboratory Practice Committee, cochairman of the Test and Technology Transfer Committee, and chairman of the Economics of Genetic Services Committee. Mr. Watson is an assistant professor of pediatrics and genetics and director of the clinical cytogenetics laboratory at the Washington University School of Medicine. As the assistant director of clinical cytogenetics at Yale University, he was involved in the development of training programs for genetics laboratory technologists and other professionals in genetic testing.

Genetic Testing and the Use of Information

ONE

Introduction

Clarisa Long

We are in the early years of a technological revolution arising from our understanding of the genetic basis of life. Scientists' ability to manipulate and decode genes is advancing with extraordinary speed. The biotechnology industry is one of the United States' most innovative sectors. The Human Genome Project, a fifteen-year multinational effort to decipher all the genetic information contained in human DNA, is proceeding faster than anticipated.

Since its development, genetic research has created high hopes for some and deep anxieties for others. Recent developments in genetic research have only multiplied the number of vexing legal, economic, ethical, and social issues. A few of those issues are: Should individuals be allowed personal property rights to their DNA, cells, or tissues? How should the industry be regulated so as to maximize safety without stifling innovation? What are the appropriate uses of gene therapy and other genetic manipulations?

The information contained in our genes is being discovered at an ever-increasing pace. Perhaps the most worrisome developments created by the combination of genetic research and information technology involve questions of privacy and genetic discrimination: Who should have access to information derived from a genetic test? Should there be an obligation to tell a spouse or a child test results? Are there times when federal, state, or local governments may appropriately mandate individual genetic testing or community-wide genetic screening? Will employers be able to require the release of genetic records as a condition of employment? Will insurance companies be allowed to use genetic information to determine risk, thus limiting the pool of insured persons so as to contain costs for the group? How will we prevent adverse selection against insurance companies by individuals? At what point does a genetic condition qualify as a disability under the Americans with Disabilities Act?

The Human Genome Project

The foundation for understanding the unique genetic profile that each of us carries is being laid by the research conducted through the Human Genome Project. The Human Genome Project is an international effort, begun in October 1990 and slated for completion in 2005, to identify all the genes in human DNA and make them accessible for further analysis. It is coordinated by the National Institutes of Health's National Human Genome Research Institute (NHGRI) and the U.S. Department of Energy. The human genome, which is the entire sum of human DNA, is estimated to contain approximately three billion subunits, or nucleotides, that make up an estimated 50,000 to 100,000 genes.

The Human Genome Project seeks to do two things: (1) to establish a map of the entire human genome, with markers positioned approximately every 100,000 bases across each chromosome, and (2) to sequence the entire human genome. One of the project's products will be a biological database that will serve as a resource for a generation or more of future research. The complete sequence of the human genome will help us understand our biochemistry, metabolic processes, neural development, aging, and a host of other factors that make us human. Large-scale automatic sequencing techniques will allow scientists quickly and easily to conduct mutation research and compare tiny variations in genetic sequences. The information generated can be used in biomedicine to understand the genetic basis of thousands of human diseases and eventually to diagnose diseases and develop optimal therapies.

Beyond providing a detailed picture of human DNA, no one can predict exactly what serendipitous discoveries lie ahead. The Human Genome Project combines the exploratory nature of basic research with the cost-benefit considerations of applied research—an important thing in these days of concern over public spending on scientific research projects. In addition to its expected benefits, the genetic information generated by the Human Genome Project carries far-reaching ethical and social implications, such as the danger of misuse and potential threats to personal privacy.

Genetic Testing

Each person inherits two copies of every gene, one from each parent. Those genes in turn can give rise directly to disorders, can create a predisposition to developing diseases, or can work with a constellation of other genes to influence one's general health.

A few genetic diseases are inherited through dominant genes. A dominant gene is one that is "expressed" or determines a specific trait or condition. Only one copy of a dominant gene is necessary for it to be expressed. Huntington's disease, which is a rare but progressive and fatal neurological disorder, is an example of a dominant-gene disorder. Thus, a person possessing even a single copy of the gene for Huntington's disease will almost surely develop the disease. Each child of a parent possessing a dominant gene has a 50 percent chance of inheriting the copy of that gene.

In the case of Huntington's disease, a genetic test has been developed to diagnose presymptomatic individuals, although no treatment has yet been developed. Extensive pre- and posttest counseling to help patients deal with the emotional and psychological issues surrounding Huntington's disease accompanies the test. Before the genetic test was developed, most (approximately 80 percent) of the people at risk for developing Huntington's disease stated that they would want to be tested when a test became available. Experience has shown, however, that in the absence of treatment, the actual percentage of people opting to take the test is much smaller.

Most of the genetic diseases we know about today are recessive disorders. A recessive condition requires two copies of a gene to be expressed. Cystic fibrosis is an example of a recessive disorder. Thus, a person who develops cystic fibrosis has two abnormal copies of the gene, one inherited from each parent. A person who has only one abnormal copy of a gene for a recessive condition is called a "carrier." While carriers will not develop the condition themselves, they have a 50 percent chance of passing on the gene to any one of their children. Children of parents who are both carriers have a 25 percent chance of inheriting both copies of the abnormal gene and developing the condition and a 50 percent chance of being a carrier.

By far the vast majority of diseases fall into a third category, the multigenic conditions, which are influenced by multiple genes, working in conjunction with environmental factors. Technologies are being developed to test for the existence of thousands of genes simultaneously, in the hopes that someday we will be able to understand how these genes interact with each other.

The genetic material for a test can be obtained from any tissue, including blood. Doctors can use genetic tests for several reasons: (1) carrier testing, or discovering whether a person is a carrier for certain genetic diseases, which involves identifying unaffected individuals carrying one copy of a gene for a disease that requires two copies for the disease to be expressed; (2) susceptibility testing, or learning whether a person has a predisposition to develop a particular disease; (3) prenatal diagnostic testing, which helps expecting parents know whether their unborn child will have a genetic disease or disorder; (4) presymptomatic testing either for predicting the occurrence of disorders such as Huntington's disease or for estimating a risk for developing a disorder such as Alzheimer's disease; (5) confirming the diagnosis of certain disorders that are initially diagnosed through other means; and (6) forensic or identity testing. The cost of a single test can reach thousands of dollars, depending on the size of the genes and the number of mutations tested.

In addition to conditions traceable to a single gene, such as Huntington's disease (caused by one copy of a single dominant gene) and cystic fibrosis (caused by two copies of a single recessive gene), genes have been implicated in predisposition to heart disease, Alzheimer's disease, diabetes, breast cancer, multiple sclerosis, the ability to mount an immune response to parasites, and other conditions ranging from the innocuous to the lethal. Such tests have so far been available mostly to families at high risk for genetic conditions or to participants in research studies. A predisposition to a disease does not mean the person will get the disease; it only means the person will have a certain increased risk of developing the disease. Environmental factors, which can lower or increase an individual's risk of developing the disease, also influence those diseases. Genetic counselors are trained in helping individuals understand the nuances of a genetic predisposition to a disease.

As this volume goes to press, tests for about 450 genetic diseases have been developed, although most of those tests are available only to research laboratories and are offered only to family members of individuals who have been diagnosed with the condition. Table 1-1 lists some of the genetic tests currently clinically available.

Hundreds of thousands of people are at risk for genetic diseases for which tests have been developed. The information provided by those tests is a mixed blessing, depending on whether the tests are carried out properly and how the information is used. Information from the burgeoning number of genetic tests will soon enable testing not just for dozens of genes that have been directly linked to serious disease and for single genes that merely predispose their carriers to certain conditions, but also for disorders that are multigenic, or due to the interaction of multiple genes.

Much of the recent debate over genetic testing has centered on commercialized tests for adult-onset diseases like Alzheimer's and some forms of cancer, like breast and ovarian cancer. Such tests only indicate a probability for developing the disorder and are targeted to healthy, presymptomatic people who are identified as being at high risk on the basis of family medical history. A positive result, being merely probabilistic, can be dif-

TABLE 1-1
A SAMPLING OF DNA-BASED GENETIC TESTS

Condition	*Test Name*	*Description of Disorder*
Alpha-1-antitrypsin deficiency	AAT	Emphysema and liver disease
Amyotrophic lateral sclerosis	ALS; Lou Gehrig's disease	Progressive motor function loss leading to paralysis and death
Alzheimer's disease[a]	APOE	Late onset variety of senile dementia
Ataxia telangiectasia	AT	Progressive brain disorder resulting in loss of muscle control and cancers
Gaucher disease	GD	Enlarged liver and spleen, bone degeneration
Inherited breast and ovarian cancer[a]	BRCA1 and BRCA2	Early onset tumors of breasts and ovaries
Hereditary non-polyposis colon cancer[a]	CA	Early onset tumors of colon and sometimes other organs
Charcot-Marie-Tooth	CMT	Loss of feeling in ends of limbs
Congenital adrenal hyperplasia	CAH	Hormone deficiency; ambiguous genitalia and male pseudohermaphroditism

(table continues)

TABLE 1-1 (*continued*)

Condition	*Test Name*	*Description of Disorder*
Cystic fibrosis	CF	Disease of lung and pancreas resulting in thick mucous accumulations and chronic infections
Duchenne muscular dystrophy/Becker muscular dystrophy	DMD	Severe/mild muscle wasting, deterioration, weakness
Dystonia	DYT	Muscle rigidity, repetitive twisting movements
Fanconi anemia, group C	FA	Anemia, leukemia, skeletal deformities
Factor V-Leiden	FVL	Bleeding disorder
Fragile X syndrome	FRAX	Leading cause of inherited mental retardation
Hemophilia A and B	HEMA and HEMB	Bleeding disorders
Huntington's disease	HD	Usually midlife onset; progressive, lethal, degenerative neurological disease
Myotonic dystrophy	MD	Progressive muscle weakness; most common form of adult muscular dystrophy
Neurofibromatosis type 1	NF1	Multiple benign nervous system tumors that can be disfiguring; cancers

Condition	*Test Name*	*Description of Disorder*
Phenylketonuria	PKU	Progressive mental retardation due to missing enzyme; correctable by diet
Adult polycystic kidney disease	APKD	Kidney failure and liver disease
Prader Willi/Angelman syndromes	PW/A	Decreased motor skills, cognitive impairment, early death
Sickle cell disease	SS	Blood cell disorder; chronic pain and infections
Spinocerebellar ataxia, type 1	SCA1	Involuntary muscle movements, reflex disorders, explosive speech
Spinal muscular atrophy	SMA	Severe, usually lethal progressive muscle wasting disorder in children
Thalassemias	THAL	Anemias—reduced red blood cell levels
Tay-Sachs disease	TS	Fatal neurological disease of early childhood; seizures, paralysis

a. Susceptibility test only; provides only an estimated risk for developing the disorder.

ficult to interpret. Many people who carry a mutation predisposing them to a condition never develop it. For any individual, it is difficult to determine the probability of developing the disease. Scientists believe that many diseases and predispositions are multigenic or are influenced by environmental factors. Those commercialized tests give rise to a plethora of questions, such as whether a particular genetic test is reliable and clinically valid; whether testing should be performed when no treatment is available; and whether parents should have the right to have their minor children tested for adult onset diseases.

The Use of Information

New genetic technologies and the information they generate are placing a new emphasis on the role of bioethics in medical and social decisionmaking. Government policymakers, lawyers, and other decisionmakers are turning their attention to genetics. Ethicists are debating whether this genetic information is the exclusive property of patients or is instead properly the concern of insurers, employers, and society. Patenting genetic products, protecting personal genetic information, preventing genetic discrimination, and developing genetic laboratory safeguards are all issues that are drawing legal and legislative attention. Nonetheless, the need for wider public understanding of the ethical, legal, and social issues surrounding genetic testing is particularly acute since many of the consequences will be fully felt only a decade or more in the future, long after the legal procedures and doctrines for dealing with genetic testing and the use of information have been established.

The American Enterprise Institute's Genetic Research and Society conference series continues to assess the technologies being created as a result of genetic research, the public policy issues surrounding genetic testing and the use of genetic information, and the ethical dilemmas created by genetic research. This volume contains essays on genetic privacy, the regulation of genetic testing, and genetic discrimination.

The past decade has seen a tremendous expansion of scientific knowledge in the field of human genetics. But much more fundamental scientific research will be necessary to put the results of genetic studies to use in the diagnosis and treatment of disease. Many of the ideas emerging from the field of biology are transforming our society—how we view our relationship with nature; how we evaluate the quality of life; and how we understand, diagnose, and treat disease. In some cases, however, not enough is known about the meaning of genetic information or how it is being interpreted and used. Transmitting genetic information from physicians to patients and vice versa will require special caution and special counseling because genetic information affects people in ways not yet fully understood. It may predict events that will occur years in the future or not at all. It may predict the future of other family members. It has a potential impact that challenges and may stigmatize both individuals and families. Genetic information will rarely provide specific directions.

Perhaps the most urgent of social issues arising from genetic research are the questions of privacy and fair use of genetic information. As David Korn notes in chapter 2 of this volume,

> If the issues surrounding genetic privacy were not a sufficient challenge to personal and social equanimity, the explosive growth of genetic knowledge and capability is taking place

> concurrently with revolutionary advances in electronic information technology and at a time when the American health care delivery system is undergoing a dramatic "industrial revolution," driven by the demand of private and public payers alike that the utilization of health care resources be rationalized and their costs contained.

Genetic research is producing information that is powerful. In many cases, such information may be potentially predictive, not just for individuals, but for families and relatives. While such information may be beneficial to individuals and their families, controlling how such information is used and who has access to it is proving problematic. Commentators have raised concern about the process of informed consent, particularly when the risks and benefits of research participation are not fully known. Others have pointed out that the existing safeguards for medical information privacy are insufficient given the increasingly computerized nature of medical practice. At the same time, the advance of science depends on continued access to stored tissue samples and the genetic information they contain. As David Korn puts it, "Since every individual is a direct beneficiary of the historic medical knowledge base, the ethical principle of distributive justice would suggest that everyone should be obligated to contribute to that base." In chapter 3 Karen Rothenberg responds that we need to consider what she calls the "three *C*s"—context, control, and community—when discussing a model that requires balancing concern for private rights with public benefit and research.

Chapter 4 of this volume deals with the regulation of genetic testing. The information generated by genetic research is changing the practice of biomedical research, clinical diagnosis of diseases, and public perceptions about genetic information and medical technology. Despite that, most health care professionals are not sufficiently knowledgeable about genetics, genetic technologies, and the ethical, legal, and social implications of having and using genetic information. Michael S. Watson argues that "medical genetics must be integrated into the broader health care delivery system rather than separated from it by genetics-specific regulations."

Watson points out that development of genetic tests has been a balancing act between the freedom of scientific research and regulation:

> [B]ecause new tests have been developed by using laboratory-based technologies and reagents rather than manufactured test kits, the role of the regulatory system has been minimized, and its development with regard to genetic testing has been slowed. An underlying premise in that regard is that oversight must avoid compromising research and investigation in deference to controlling the practice of medicine.

One of the limitations of all medical tests, whether genetic or not, is the possibility of laboratory errors. Such errors might arise from misidentifying samples, contaminating the test materials, improperly administering the test, or other sources. Watson concludes that "significant improvements in the use of genetic testing will come from education of the public and health care providers and the maximal acquisition of data from providers of genetic testing services and researchers to guide our decisions about appropriate, safe, and effective use."

Chapters 5 and 6 present essays on genetic discrimination. Some scientists and physicians believe that while personal knowledge of genetic information can sometimes benefit an individual by allowing for early, more accurate, and effective treatment, the uncertainties surrounding the interpretation of genetic tests, personal anxiety, and social stigmatization could outweigh the benefits of testing. They claim that genetic testing carries with it a significant threat of discrimination by potential employers and insurers.

Philip R. Reilly disagrees. He argues that while a widespread fear exists that genetic testing carries with it a significant risk of genetic discrimination, "little evidence exists to support the widespread fear that people who undergo genetic tests to determine whether they are at increased risk for developing a serious disorder face a significant risk of genetic discrimination." He contends that while state laws have provided little actual protection against such discrimination, federal laws have "sharply reduced the risk of genetic discrimination in regard to access to health care and employment."

Ellen Wright Clayton disagrees with Reilly's conclusions that the risk of genetic discrimination is low. Instead, she points to other reasons why documented cases of genetic discrimination are few—the empirical studies to date on genetic discrimination may have been flawed and our understanding of genetic tests is in its infancy—and remains skeptical of the actual effect federal legislation will have, in the long run, in preventing genetic discrimination.

These chapters collectively seek to address some of the major issues confronting a society on the brink of a new genetic information age. Every society must find a way to answer the basic question, Who are we? As the ever-increasing torrent of genetic information becomes a deluge, that question is presented on an increasingly individual level. Will our society accept as true what Nobel Prize winner James Watson said when he headed up the Center for Human Genome Research at the National Institutes of Health: "We used to think our fate is in our stars. Now we know, in large measure, our fate is in our genes." To what degree is biology destiny? To what extent are human talents and institutions the product of our anatomical, neurological, hormonal, or genetic constitution? Our society has gone through pendulum swings that over- or undercorrelate DNA with behavior and disease. We have much to learn about the complex interactions within an organism and between an organism and its environment that influence its development. In the meantime, we must decide the legal and ethical role we will accord genetic information as genetic research presents us with those increasingly pressing and inescapable questions.

TWO

Genetic Privacy, Medical Information Privacy, and the Use of Human Tissue Specimens in Research

David Korn

Dramatic advances in basic science and technology, significantly stimulated by the Human Genome Project, have led to remarkable progress in molecular genetics and our understanding of normal and abnormal genomic structures. Those advances have already transformed biomedical research and promise in coming decades to have an equally profound impact on the practice of medicine, including not only diagnosis and therapy, but strategies of disease prevention and health promotion. At the same time, however, those accomplishments raise a myriad of difficult policy issues with which society has only begun to grapple. In particular, those accomplishments have generated deep social concerns about the acquisition, protection, and use of *genetic information.* That term, freighted with mystique and imperfectly understood by most of the populace, is generally regarded with awe and fear: awe because the information is perceived to be intensely personal, private, powerful, pedigree-related, and predictive[1] and fear because the potential misuse of such information can lead to insurance and employment discrimination, disruption of personal and familial well-being, and stigmatization.

If the issues surrounding genetic privacy were not a sufficient challenge to personal and social equanimity, the explosive growth of genetic knowledge and capability is taking place concurrently with revolutionary advances in electronic information technology and at a time when the American health care delivery system is undergoing a dramatic "industrial revolution," driven by the demand of private and public payers alike that the utilization of health care resources be rationalized and their costs contained. Those changes in themselves have generated widespread public unease about the erosion of individual privacy and the security of sensitive personal information of all kinds. The pervasive unease becomes heightened with respect to medical information, where the privacy issues raised by all three areas of rapid change converge, and the unease approaches alarm when that information bears the iconic label "genetic" (Nelkin and Lindee 1995).

The policy issues surrounding the privacy of medical information in general, and genetic information more specifically, are challenging, and their intimate interlinkage only adds to their complexity. The issues are suffused with symbolic meaning and emotion that confound dispassionate public discourse. They are very difficult, for they lie at the intersection of patient diagnosis and management, biomedical research, and pub-

1. I first heard this description in a talk given by Dr. Francis Collins, director of the National Institute of Human Genome Research, who referred to genetic information as being characterized by the five *P*s.

lic health, and they require adjudication of important, but competing, private and public goods. Given that we are still at the dawn of both the new genetics and information ages and considering that the profound changes transforming the American system of health care delivery and financing are themselves likely to represent only a transitional stage whose ultimate outcome is still unknown, the parameters that are established now to guide the public policy debate may be as important as the specific legislative and regulatory solutions that come forth. All the questions cannot be answered now, nor even asked, and society will surely be forced to revisit those matters repeatedly as the foundational sciences and technologies continue to progress and public understanding and sentiment continue to evolve.

During the past several years, public concern about privacy and the security of personal information has tended to focus sharply and narrowly on genetic information and, more specifically, on the nature of the informed-consent process that should guide genetic testing and the accessibility of human tissue samples for biomedical research. Only within the past two years have the broader and in many ways more complicated issues of the privacy and security of medical information begun to receive equal or greater attention, especially in Congress. Both those topics have generated a substantial literature as well as considerable legislative activity at both the state and federal levels. In addition, medical information privacy has been the subject of several formal reports from the Department of Health and Human Services (1995), the Institute of Medicine (1997), the National Research Council (1997), and the Office of Technology Assessment (1993).

The general literature on medical and genetic privacy (see, for example, de Gorgey 1990; Gostin 1995a,

1995b, 1997; Gostin et al. 1996; Powers 1994; Reilly, Boshar, and Holtzman 1997; Westin 1976) has been authored predominantly by medical ethicists, bioethicists, and lawyers, who have tended to approach the issues almost exclusively from the perspective of individual privacy rights and whose arguments spring from such common ground as the Nuremberg Code, the *Belmont Report* (National Commission for the Protection of Human Subjects of Biomedical and Behavioral Research 1997), the federal Privacy Act of 1974, and federal regulations governing research on human subjects, which are frequently referred to as "the common rule" (Federal Policy for the Protection of Human Subjects 1991). The specific issue of the use of human tissue samples in research has been rather intensively addressed in the past few years in a number of position papers, pronouncements, and recommendations from committees and working groups, some sponsored by the Ethical, Legal, and Social Implications (ELSI) Program of the Human Genome Project, others representing genetics societies and patient advocacy groups (see, for example, American College of Medical Genetics 1995; American Society of Human Genetics 1996; Clayton 1997; and Clayton et al. 1995). Those contributions differ somewhat both in their definition and assessment of the problems and in their specific recommendations for regulatory or legislative remediation. Not surprisingly in light of their authorship, all of them reflect the heavy input of bioethical, legal, and patient-privacy perspectives and are marked by a notable deficiency of input from the broad scientific community.

In fact, only within the past three years have those issues finally begun to draw attention from the wider community of biomedical scientists, where the proposals have raised considerable distress, especially among those who are increasingly applying powerful new technologies to human tissue samples to study diseases of major social import. Those investigators recognize that the tools of contemporary molecular biology and genetics have brought to such research a heretofore unimagined capability to gain unique insights into human diseases, and they fear that the restrictions and proscriptions called for in those proposals would severely burden, if not fatally encumber, an entire class of promising research.

Recent activity in the legislative arena was represented by a number of bills that were introduced in the 105th Congress, some of which have now been reintroduced in the 106th Congress, and by laws that many states have already enacted. Some of the congressional bills under consideration in the spring of 1999 that attempt to tackle the broad issues of the privacy and security of medical information and records are H.R. 1057, Markey; S. 578, Jeffords-Dodd; S. 573, Leahy-Kennedy; and S. 881, Bennett. Several other bills primarily concerned with what is often called "patient rights" or "patient protections" (S.300, Jeffords; S. 326, Lott; S. 240, Daschle; H.R. 358, Dingell; H.R. 448, Bilirakis) or "managed care reform" (H.R. 719, Ganske; S. 374, Chafee) incorporate provisions of variable length and complexity that deal with medical information privacy.

Other bills introduced in the Congress in the past few years focused more narrowly on protecting genetic privacy and preventing the discriminatory misuse of genetic information, primarily in the health insurance markets and employment, but sometimes, as well, in the disability and life insurance markets (S. 422, Domenici; H.R. 341, Stearns; H.R. 306, Slaughter; S. 89, Snowe; H.R. 2215 and 2216, J. Kennedy; S. 1045, Daschle; S. 1600, Feinstein; H.R. 4008, Solomon). In addition, the Health

Insurance Portability and Accountability Act of 1996 provides limited protection from genetic discrimination to individuals covered by group health insurance plans and mandates that Congress pass comprehensive medical information privacy legislation by August 1999, failing which, the secretary of health and human services must promulgate regulations.

It is noteworthy that to date the Clinton administration has refrained from proffering its own legislation either on medical or genetic information privacy or on genetic discrimination. Secretary Donna Shalala of the Department of Health and Human Services did, however, submit to the president in July 1997 a report, *Health Insurance in the Age of Genetics,* containing her recommendations for federal antidiscrimination legislation. She also submitted a report to Congress in September 1997 entitled *Confidentiality of Individually Identifiable Health Information* that presented her recommendations of the overarching principles (but few details) that should inform federal legislation on that subject. The latter report, as required by the Health Insurance Portability and Accountability Act, reflected the strong input of the National Committee on Vital and Health Statistics (NCVHS 1997), whose recommendations on health information privacy were delivered to the secretary in June 1997. Finally, the National Bioethics Advisory Commission, which at the request of the president has been studying the use of stored human tissue samples in research since its inception in October 1995, issued in January 1999 for review and comment a draft report that called for new legislative and regulatory initiatives. A final report is expected in the summer of 1999.

As of June 1998, at least twenty-four states had enacted anti–genetic discrimination laws, and all other states had introduced bills. The legislative efforts by the states have tended to focus narrowly on the issues of genetic discrimination (Rothenberg 1995; Rothenberg et al. 1997). More often, they deal with discrimination exclusively in health insurance; less often, they deal in addition with discrimination in employment. A noteworthy exception is Minnesota, which in January 1997 implemented a statute (Statute 144.335) that extends broad protection to health records and prohibits any use of them in research without the express written authorization of the patient. Although the remarkable volume of state legislative activity clearly expresses the high degree of public concern about genetic discrimination, especially in health insurance, the statutes and bills frequently seem to represent hasty responses to public pressure and emotion. They tend to be confounded and often truncated in their reach by definitional ambiguities involving such terms as *genetic testing, characteristic, information, makeup, anomaly, predisposition,* and *defect,* as well as by the statutory preemption from state insurance laws granted employer-sponsored, self-funded health plans under the Employee Retirement Income Security Act. In addition, they often bear the scars of heavy lobbying by the insurance industry, which to date has been largely successful in preventing the enactment of protections against discrimination on the basis of a genetic predisposition in life insurance. As a result, those efforts have yielded an uncoordinated and often discordant legislative patchwork of uneven scope and effectiveness.

The federal regulatory scene is not much clearer. For example, in 1991 the Equal Employment Opportunity Commission opined that the Americans with Disabilities Act did not protect individuals from discrimination based on genotype alone. But in 1995 the commission did extend partial protection from genetic discrimination in employment by issuing an interpretive guideline stating that a person denied employment on the basis of a disease-susceptibility gene would be covered by the act. To my knowledge, that ruling has not yet been tested in court.

Finally, no review of legislative activities in the complex arena of medical information privacy in an age of medical information technology can be complete without mention of the European Union Data Privacy Directive adopted in October 1995 and the Council of Europe's committee of ministers' "Recommendations on the Protection of Medical Data," adopted in February 1997. Lowrance, in his excellent May 1997 report to Secretary Shalala entitled *Privacy and Health Research,* described the directive as a "framework directive" establishing general principles, with which the fifteen EU member states had to bring their national laws, regulations, and administrative provisions into congruence by October 1998. In spring 1999 that process was still underway. Lowrance also noted that it was still not clear how the stringent requirements of the EU directive and the equally stringent provisions of the committee of ministers' recommendation would be made to mesh with one another. Those requirements have aroused deep concerns within the European community of medical and health services researchers, who believe that, unless very carefully implemented by each nation, the requirements will have a chilling effect on the large body of medical, epidemiological, and health services research that is totally dependent on access to archived medical records (Knox 1992; Lynge 1994; Vandenbroucke 1998; "Pharmacoepidemiologists Moving" 1997). Both the directive and the recommendations impose severe restrictions on the processing or transmission of "personal data"—by which is meant "any information relating to an identified or *identifiable* natural person" (emphasis added)—and in particular prohibit the transfer of personal data to a non-EU nation unless the recipient country "ensures an adequate level of protection." That provision, which will shortly be embedded in European privacy protection statutes and policies, is of immediate consequence to the United States and other non-European nations and has been a powerful force driving the perceived need for rapid enactment of strong, comprehensive federal legislation on the privacy of medical information.

Although the theme of this chapter is the use of human tissue samples in research, I have chosen to provide this brief summary of commentary and legislative activities relating to the privacy of both medical and genetic information because of my conviction (to which I return later in this chapter) that feasible solutions to the problems of access to and protection of both kinds of information (in whatever form) are largely indistinguishable. Put differently, the two kinds of information are in fact so intimately intertwined that they cannot be segregated legislatively or by regulation in any way that would prove operationally feasible. Accordingly, the content and fate of the several medical information privacy bills now pending in Congress are of immediate (and perhaps, controlling) relevance to the ultimate management of genetic information privacy and access. It is also pertinent to note that with respect to medical privacy legislation, and in sharp contrast to the deliberations to date about genetic privacy, the broad scientific community—well aware of the crucial importance of access to archival medical information for medical, epidemiological, public health, and health services research—has been vigorously engaged in attempting to inform and shape the congressional bills from the outset. The impact of their participation has been telling.

The scientific community's expression, albeit belated, of concerns about the direction and content of the public discourse to date about genetic privacy and access to human tissue samples for research strongly argues for the need to broaden participation in the dialogue to arrive at understandings and principles that can lead to generally acceptable proposals and workable guidelines. In the remainder of this chapter, I provide some background and context to clarify and help frame the issues, review some of the most worrisome features of the recommendations about tissue access that have been proposed, and make some general observations about the public deliberative processes to date. Following that, I present a number of specific recommendations that are intended to stimulate the dialogue and help bring both scientific perspective and reasonableness to the difficult task of striking a proper balance between the important competing claims of private interest and public benefit. In presenting the recommendations, I relate them wherever possible to the continuing discourse about the privacy of medical information, toward which Congress has currently turned its attention, because the outcome of that discourse will most likely determine our nation's course on such very difficult issues.

Background and Context

The history of medical progress is deeply rooted in the careful study of archival patient materials—medical records and tissue samples. Our nation's hospitals, especially the academic medical centers, collectively contain an enormous archive of human tissue samples, which constitute a unique resource that records the prevalence and protean expressions of human diseases over time. The tissue specimens were removed for medical reasons, under sparing consent language that usually included a proviso for research and educational uses, and were submitted to the pathology laboratory for routine diagnostic evaluation, after which portions of the specimens were permanently stored as part of the patient's medical record in accordance with sound medical practice and legal and accreditation requirements. Although not collected for research purposes, those specimens have served as a rich source of materials for clinicopathological investigations and yielded a wealth of insights into human diseases that have provided over more than a century most of the vocabulary and much of the foundation of modern medicine (Korn 1998a). Yet, because of the limits of the available technology, the results of the studies have historically been primarily of public benefit and of little immediate consequence to individual patients (or "sources") from whom the tissues had been obtained. Accordingly, the practice and standard of informed consent for that vast body of research has traditionally been quite minimal.

What has now changed dramatically has been the introduction into practice of powerful new techniques like monoclonal antibodies and the polymerase chain reaction, which pathologists and other investigators can apply in the research laboratory to fixed, paraffin-embedded, and even sectioned tissue specimens—not only to demonstrate specific abnormalities of gene structure and expression, but often to infer whether the changes are of somatic origin or present in the source's germline DNA and, therefore, hereditable. The power of those approaches provides unique insights into the mechanisms of human diseases (for example, neoplasms) that cannot be obtained by other means; and they offer the promise of supporting major advancements in diagnosis, prognosis, therapy, and even prevention. At the same time, however, by their nature those results may be seen as constituting unauthorized probing into the most intensely intimate recesses of an individual's self, and they may on occasion be construed to have major predictive consequences not only for the individual patient sources but also for their families. The emergence of those technological capabilities has wrested the entire topic of research on human tissue samples from its historic state of repose in obscurity and thrust it sharply into the consciousness of a public already uneasy about erosion of privacy, fearful of discriminatory misuse of personal health information, and susceptible to being stirred up by anything containing the iconic words *gene* or *genetic.*

The oversight of research involving human subjects in the United States has been an evolving process since the end of World War II, and the conceptual path is lighted by such landmark documents as the Nuremberg Code of 1947 and the Helsinki Declaration of 1964 (World Medical Association 1989) and by the 1979 *Belmont Report.* That report laid down the three basic ethical principles that undergird the protection of human subjects in research: respect for persons (to treat persons as autonomous agents and grant special protection to those whose autonomy is reduced); beneficence (to maximize possible benefits and minimize possible harm); and justice (to distribute the benefits and burdens of research fairly). Regulations dealing with the oversight of human subjects research by the Department of Health and Human Services were first codified in 1974 in the Code of Federal Regulations (45 C.F.R. 46) and have been frequently revised. In 1991 the core provisions of 45 C.F.R. 46 were incorporated into a model Federal Policy for the Protection of Human Subjects and adopted by sixteen additional federal departments and agencies that sponsor, conduct, or regulate research involving human subjects. The Food and Drug Administration had and continues to maintain its own variant of that policy. The "common-rule" policy is overseen by the Office for Protection from Research Risks (OPRR, an entity of the Department of Health and Human Services but housed within the National Institutes of Health) through locally appointed bodies called institutional review boards (IRBs), which must review and approve all proposals for federally sponsored research involving human subjects before that research can be funded and proceed. Although many institutions that conduct federally sponsored research have elected by institutional policy to extend the purview of their IRBs to include *all* human subjects research irrespective of source of funding, the present protections of the common rule are incomplete. To remedy that deficiency, Senator John Glenn introduced into the 105th Congress a bill (S. 193) that would extend the common-rule protections to all human subjects research, irrespective of research sponsor, and would strengthen and relocate the functions of OPRR to the Office of the Secretary of the Department of Health and Human Services. The bill, heatedly opposed by industry, languished and was not pursued by that Congress.

Human subject is defined as a living individual about whom an investigator conducting research obtains data from intervention or interaction with the individual or from individually identifiable private information, an example of which is the medical record. Although the definition is broad, several categories of research are specifically exempted from the regulations. They include

> research involving . . . the study of existing data, documents, records, pathological specimens or diagnostic specimens, if these sources are publicly available or if the information is

> recorded by the investigator in such a manner that subjects cannot be identified, directly or through identifiers linked to the subjects.

The federal regulations deal extensively with the elements of information and choice that constitute appropriate informed consent. The requirements are many, and the procedure is an elaborate one, which, although appropriate in the typical clinical research setting involving direct interaction of researchers with subjects, has not historically been considered necessary by researchers or IRBs with respect to research on tissue samples obtained in the course of ordinary medical care.

Accordingly, the consent forms routinely used for hospital admissions or surgical procedures, particularly in academic medical centers, have typically contained a simple line or two within the text stating that any removed tissue not required for diagnostic purposes may be used for teaching and research. Under the regulations, an IRB can approve such an informed consent, which I refer to as a version (admittedly minimal) of *general* informed consent—to be contrasted with *specific* informed consent, which is much more detailed and stringent. An IRB may also approve the absence of informed consent under a waiver provision, if it finds that: "(1) the research involves no more than minimal risk . . . ; (2) the waiver . . . will not adversely affect the rights and welfare of the subjects; and (3) the research could not practicably be carried out without the waiver." (Note that the key terms in the waiver provision are *minimal risk* and *practicably.*) Finally, the regulations also provide that noninterventional research activities of no more than minimal risk involving the study of existing data, documents, records, pathologic specimens, or diagnostic specimens may be reviewed by the IRB under an abbreviated, expedited procedure. In 1997 both the OPRR and the FDA issued companion notices of proposed rulemaking ("Protection of Human Subjects," 62 *Federal Register* 60,607, 1997), recommending, among other changes, a relaxation of the language that defines the class of studies eligible for expedited review as follows:

> Research involving existing identifiable data, documents, records, or biological specimens (including pathologic or diagnostic specimens) where these materials in their entirety have been collected prior to the research for a purpose other than the proposed research; [and] Research involving solely (a) prospectively collected identifiable residual or discarded specimens, or (b) prospectively collected identifiable data, documents, or records, where (a) and (b) have been generated for nonresearch purposes.

In the final rule (63 *Federal Register* 6033, 1998) that language was modified to read:

> Research involving materials (data, documents, records, or specimens) that have been collected, or will be collected, solely for nonresearch purposes (such as medical treatment or diagnosis).

An explanatory note follows to indicate that the listing refers only to research that is not exempt. That change, which apparently received overwhelming support from those responding to the Notice of Proposed Rulemaking, would appear to fly in the face of the sundry proposals calling for regulatory tightening of access to stored human tissue samples for purposes of genetic research, a topic that I review in the next part of this chapter.

On the basis of those several provisions in the common rule, and especially on the assumption of "no more than minimal risk" and "impracticability," clinicopathological research on archival human tissue specimens—whether or not directly identified or coded and therefore linkable to individual patients—has been conducted for decades either without or under waived or expedited IRB review and under a standard of informed consent that, although de minimis, has been generally deemed sufficient.

The ability of modern molecular technology to extract precise genetic information from any tissue sample that contains DNA has generated widespread anxiety about the potential uses and abuses of such information. That anxiety resonates with and is amplified by a deeper and more pervasive public angst about privacy and confidentiality in a clinical care environment increasingly marked by electronic medical records and managed care practices. As noted earlier, those concerns have stimulated numerous federal and state legislative initiatives aimed at curbing the discriminatory misuse of genetic information as well as the formation of numerous committees to take fresh looks at the protection of individual autonomy and privacy in research and at the requirements for informed consent in genetic testing and research with human tissue specimens.

Review of Recommended Proposals

Although the National Human Genome Research Institute (NHGRI, formerly the National Human Genome Research Center) and its ELSI Working Group promised at the outset to be particularly influential in the debate, the NHGRI and the NIH have consistently refrained from issuing an official opinion or formal statement. The reticence appears not to have been unintentional and was an important consideration in the president's decision to highlight that issue as one of the first two tasks identified in his charge to the freshly convened National Bioethics Advisory Commission. But a number of position papers, some sponsored by the ELSI program, have been published by organizations such as the American College of Medical Genetics (ACMG 1995), the American Society of Human Genetics (ASHG 1996), the Task Force on Genetic Testing (1997b), a committee working under the auspices of the National Action Plan for Breast Cancer, and a working group convened jointly by the NIH and Centers for Disease Control and Prevention (Clayton et al. 1995). In addition, a document entitled *The Genetic Privacy Act and Commentary* (Annas, Glantz, and Roche 1995b), an ELSI-sponsored work product that proposes sweeping model legislation, has been widely publicized and circulated in Congress and the state legislatures, where it has proved highly influential.

Many of the central concepts and provisions of the proposed Genetic Privacy Act (GPA) are reflected in state bills that are, or were recently, pending (Louisiana, Missouri, and Nebraska) or have already been embedded in legislation (Oregon Statutes 659.036, 659.277, 659.700, 746.135 1995; New York Civil Rights Law A79-L 1996). In two other states (New Jersey and New Mexico) bills containing provisions of the proposed GPA were vetoed. In New Jersey, Governor Christine Whitman's veto was specifically directed at a provision of the bill that came straight out of the proposed GPA and would have recognized an individual's property right in excised tissue samples. (Subsequently, the New Jersey legislature passed a revised bill without the offending provision, which has been signed into law.) Moreover, the proposed GPA has strongly influenced the many drafts of the

Domenici bill (the Genetic Confidentiality and Nondiscrimination Act) that were repetitively introduced and withdrawn for further revision in the 104th and 105th sessions of the Senate, as well as the McDermott bill (H.R. 1815, the Medical Privacy in the Age of New Technologies Act of 1997).

Finally, the OPRR, whose role in interpreting federal policy and overseeing the IRB process gives it special authority in that area, took notice of some of the newly emerging issues of genetic privacy by issuing in 1993 a new chapter entitled "Human Genetic Research" for the *Institutional Review Board Guidebook* (OPRR 1993), which is the gold standard in that field. The chapter points out that contemporary genetic research raises new ethical issues and concerns "for which no clear guidance can be given at this point, either because not enough is known about the risks presented by the research, or because no consensus on the appropriate resolution of the problem yet exists." The document goes on to emphasize that the nature of the risks presented by genetic research protocols is different in that the protocols involve considerations of psychosocial, and not physical, harm, and that, accordingly, such protocols, including by inference those involving archival tissue samples and existing records, should no longer automatically be considered of "minimal risk" and thereby eligible for expedited (or waived) IRB review.

It is useful to examine some of the general features of those proposals and to look closely at two of them: *The Genetic Privacy Act and Commentary* (Annas, Glantz, and Roche 1995b), which is the most extreme of the documents but reflects an important aspect of bioethical and legal thought on those matters; and the consensus statement from the NIH-CDC Working Group (Clayton et al. 1995), which, although attempting to strike a posture of moderation, has nonetheless raised serious misgivings among many in the scientific community.

All the proposals of concern rest on the explicit proposition that genetic information is qualitatively different from all other types of personal (or medical) information in ways that require special protection and on a second proposition, whether explicit or implicit, that individuals have property interests in their DNA samples. On those propositions are built the arguments calling for restriction and proscription of sample use and for mandatory protocols of stringent informed consent for research involving the use of human tissue samples. Thus, Annas, Glantz, and Roche (1995b) argue that

> no stranger should have or control identifiable DNA samples or genetic information about an individual unless that individual specifically authorizes the collection of DNA samples for the purpose of genetic analysis, authorizes the creation of that private information, and has access to and control over the dissemination of that information.

Under the proposed GPA each person who collects a DNA sample for the purpose of performing genetic analysis must, among other things, provide a notice of rights and assurances before the collection of the sample, obtain written authorization, restrict access to DNA samples to persons authorized by the sample source, and abide by the source's instructions regarding the maintenance and destruction of the sample. The required authorization contains many specific stipulations, including a description of all approved uses of the DNA sample, whether or not the sample may be stored in an identifiable form, whether or not the sample may be used for research under any circumstances and, if so, what kinds of research are allowed or prohibited, and whether or not and with whom the sample may be shared. The proposed act states that an individually identifiable DNA sample is the property of the sample source, who thereby retains the right to order or delegate the authority to order its destruction at any time. In any event, such a sample must be destroyed upon the completion of the specifically authorized genetic analysis (analyses) unless retention has been explicitly authorized or all identifiers are destroyed. In other words, the authorization is case- or project-specific, time-limited, and revocable without limitation.

It is noteworthy that nowhere do Annas, Glantz, and Roche (1995b) explicitly define *genetic test* or *genetic analysis.* What they define is *private genetic information:*

> any information about an identifiable individual that is derived from the presence, absence, alteration or mutation of a gene or genes, or the presence or absence of a specific DNA marker or markers, and which has been obtained from an analysis of the DNA of an individual or of a related person.

In their accompanying commentary and elsewhere (Annas, Glantz, and Roche 1995a), the authors acknowledge that there are many different ways to derive genetic information, even "private genetic information," in the clinical setting, beginning with the medical and family history and extending to numerous clinical laboratory tests that may be performed in the course of a routine medical examination. The authors note, however, that distinguishing between "private genetic information" and other medical information derived from a family history would be problematic, and including all such information in the definition would make virtually all medical records subject to the provisions of the act and require the overhaul of "well-established medical information practices and procedures." From a similar analysis of common biochemical tests, they arrive at the same conclusion—to exclude them: "We have opted to exclude this type of genetic information to avoid the enormous practical problems presented by including it." The discussion clearly acknowledges the authors' discomfort over the artificiality of their proposal to carve out "private genetic information" and the impracticality of trying to impress all possible forms of such information within their definition. It is unfortunate that they did not extend their quite thoughtful analysis to the entirely analogous impracticalities that arise from attempting to apply their definition for statutory or regulatory purposes to contemporary medical diagnostic and research practices. Although several distinguished bioethicists (Buchanan 1998; Murray 1997; Reilly 1995) have thoughtfully and persuasively rebutted the proposed GPA's central findings and conclusions, the document has been highly effective in shaping legislative initiatives at both the state and federal levels.

The various proposals are in broad agreement that informed consent procedures for "genetic research" on human tissue samples should be significantly strengthened and that the standard of informed consent may vary depending on whether studies are retrospective or prospective and whether the samples to be used are anonymous or not; but the proposals differ in their details. For example, the report from the ASHG (1996) states that all human genetic research protocols must undergo IRB review and rather elaborately stratifies the stringency of informed consent required by whether the studies are retrospective or prospective and whether the tissue samples are anonymous, anonymized (originally identifiable but stripped irreversibly of all identifiers and impossible to link to sources), identifiable (linkable to sources, as through the use of a code), or identified. The ASHG's report argues that stringent informed consent is necessary for *all* prospective studies and for retrospective studies with identified or identifiable samples, but not with anonymous or anonymized samples. It further states that "it is inappropriate to ask a subject to grant blanket consent for all future unspecified genetic research projects on any disease or in any area if the samples are identifiable in those subsequent studies," an opinion shared by many. Moreover, in retrospective studies with identifiable samples, consent by recontact of sources should generally be required, although investigators may seek IRB waivers of such a requirement.

The terms *genetic tests, genetics research,* and *genetic information* are sometimes used in the various publications without further definition. Where they are defined, however, they are done so very broadly. A common, popular definition of *genetic test*—the definition used in the bill introduced in the 105th Congress by Representative Stearns (H.R. 341, the Genetic Privacy and Nondiscrimination Act) and in the aforementioned Oregon statute—is "a test for determining the presence or absence of genetic characteristics in an individual, including tests of nucleic acids such as DNA, RNA and mitochondrial DNA, chromosomes or proteins in order to diagnose a genetic characteristic." In the report from the Task Force on Genetic Testing (1997a) *genetic test* is defined as "the analysis of DNA, RNA, chromosomes, and certain metabolites [proteins or other gene products] in order to detect heritable disease-related genotypes, mutations, phenotypes, or karyotypes for clinical purposes." It is remarkable that among all the documents produced to date, only the task force report (the final version of which was very responsive to comments received on an earlier circulated draft) strikes a blow for semantic precision and clarity of thought by explicitly excluding from its definition genetic tests conducted purely for research, as well as tests for nonheritable (somatic), as opposed to heritable (germline), mutations. Indeed, none of the other documents even takes notice of, let alone attempts to operationalize, the distinction between somatic and germline mutations or incorporates into its regulatory proposals the fact that only the latter are (probabilistically) predictive and pedigree-related and, therefore, of concern to kin.

The consensus report from the NIH-CDC Working Group (Clayton et al. 1995), which has been widely cited, adopts a stringent position on most of those issues. It narrowly interprets the provisions of the existing regulations and states that all research with human tissue samples—whether anonymous, anonymized, or not—requires IRB review and that stringent informed consent should be required for tissue samples that are obtained during the course of clinical care. Some members of the committee argued that even preexisting samples should not be anonymized without explicit consent and that consent should be required with prospective tissue samples even if the samples are to be used under conditions of strict anonymity. With respect to the existing criteria for IRB waivers, the committee argued that the threshold for a burdensome or impractical requirement has not been tested but should be very high.

The committee recommendations stated that specific consent should always be required for research with linkable specimens and that "rarely does the language in typical operative and hospital admission consent forms provide an adequate basis for inferring consent to genetic research." With respect to making existing identifiable samples anonymous for use in research, the IRB should determine whether the data sought could be obtained by a different protocol that would allow for informed consent,

> whether the proposed investigation is scientifically sound and fulfills important needs[,] . . . and how difficult it would be to recontact subjects (or, by implication, next of kin). *Even with respect to research on archival, anonymous samples, the committee argued that* [emphasis added] IRBs . . . could usefully review such protocols to determine whether they are scientifically sound . . . whether they propose to address a significant problem, and whether the desired information could be obtained in a protocol that allows individuals to consent. (Clayton et al. 1995, 1791)

The last recommendations are especially troubling in that they would not only substantially expand the authority of IRBs to challenge scientific protocols in evaluating scientific merit, but would also introduce into the review process in a very explicit way a requirement to "balance values" (for example, the significance of the problem, the fulfillment of important needs versus the subject's right to privacy or, more narrowly, to genetic privacy). Such requirements would lay on those already heavily burdened and typically voluntary bodies an increased workload and responsibilities that might well exceed their competencies. Moreover, in academic settings, in which the majority of the research proposals have already passed successfully through a process of merit evaluation by peers, such an expanded scope of IRB authority would likely raise serious questions of infringement of the academic freedom of faculty investigators under prevailing university policies. Even more worrisome, any mandate requiring IRBs to weigh "competing values" in assessing research proposals would force those bodies to render judgments about research proposals for which there can be no clear-cut or normative standards and would make the decisions of the more than 3,500 IRBs estimated to exist in the United States at this time vulnerable to the personal belief structures, biases, and ideologies of their more than 17,500 individual members.

General Observations about the Process to Date

The public discourse to date and the products it has generated have been misdirected and disappointing in at least four major respects. First, it is extraordinary and saddening that in attempting to deal with issues that largely center on preventing the misuse of genetic information developed in research, so little attention or creative thought has been given to strengthening the protection of that information, while so much effort has been directed at burdening its acquisition by encumbering the conduct of genetic inquiry and erecting complicated new barriers to the continuing creation of the knowledge base.

Second, it has been too readily taken as given that genetic information is unique and different in kind from all other forms of private, sensitive, and often predictive information that may exist in a medical record—a position that has been dubbed "genetic exceptionalism" (Murray 1997). To the contrary, I would argue strongly, as have others (Gostin 1995a, 1995b; Lowrance 1997; Murray 1997; NIH 1993) that the distinction is fallacious and that efforts to operationalize the distinction must inevitably fail. That becomes particularly clear when attempts are made to separate genetic information from all other medical information present in a clinical or research record for purposes of restriction, protection, or regulation. The concept of uniqueness is problematic for another reason: it provides much of the conceptual underpinning for the proposition that patients or sources have enduring property rights in their excised tissue samples and thereby undergirds the more vexing provisions of the stringent consent protocols that have been proposed for research on those samples.

Third, the discourse has contributed to semantic ambiguity and confusion of thought by failing to distinguish between two quite different issues: genetic testing, which raises important considerations of definition and appropriate informed consent, and genetic information, which can be obtained or inferred from a myriad of clinical and research sources and which raises concerns of security and protection of confidentiality.

Fourth, until very recently there had been far too little engagement with the broad community of scientists—not only molecular scientists who work with human tissue samples, but other health researchers for whom archival patient records are the primary research materials. The absence of the latter largely results from the deception produced by framing the debate in terms of genetic exceptionalism, which has led other health researchers until very recently to believe that legislative and regulatory restrictions on access to genetic information would not affect their ability to access medical records. As I previously noted, the public discourse has reflected an abundance of bioethical and legal perspectives, but it has been insufficiently informed of contemporary medical and biomedical research practices and has been insensitive to scientific opportunity. Accordingly, the proposals offered have come down excessively heavily on the side of private interest at the expense of public benefit and thus have distorted the delicate equipoise that must always be respected in research involving human subjects.

Specific Recommendations

In this section I present eight specific recommendations for resolving issues raised in the policy debate about genetic privacy, the privacy of medical information, and the use of human tissue specimens in research.

Defining Terms Carefully. Much more careful attention must be paid to the definition of terms to enhance public understanding, to sharpen the focus and precision of public discourse, and to avoid both confusion and unintended adverse consequences. The absence of commonly accepted, practicable definitions is troublesome. Terms like *genetic research, genetic sample, genetic test,* and *genetic information* have been used inconsistently and often inappropriately with reference to modern clinical or research practices. In the context of contemporary medical practice and biomedical research, those terms become so inclusive and imprecise that it is inadvisable and impractical to attempt to use them without very careful circumscription as the basis for new research guidelines or regulations.

Genetic testing, appropriately defined, should unarguably meet a high standard of informed consent. The definition of a genetic test, however, should be narrowed to focus on the purpose of the study rather than on the particular kinds of research methodologies to be used. For example, such a test might be one that is carried out on asymptomatic individuals or on populations to determine the presence or distribution of particular heritable risk factors of established predictive significance for purposes of genetic counseling, medical management, or epidemiological or public health surveys. Note that I would explicitly exclude from the definition all tests of established utility that are carried out for diagnostic purposes in the course of providing medical care. The results of such tests become part of the medical record and must not be encumbered with special restrictions beyond those appropriate for protecting the confidentiality of all medical information created in the clinical setting.

Research studies on human tissues removed for medical reasons ordinarily should not be defined as genetic tests, even if they involve examination of gene structure and function. Research results are nascent data, obtained under experimental conditions that would not usually meet accreditation standards for diagnostic clinical laboratories. Research results cannot be fully interpreted until they have been adequately replicated, and only after they have been validated analytically and clinically can the medical community determine whether they provide the basis for a useful genetic test that should be introduced into practice. Accordingly, research results as a rule should not be considered diagnostic and should not be entered into a medical record or communicated directly to a tissue source. In the event that research yields information that the investigators believe could be of significant consequence to particular individuals, the investigators should bring the matter before the cognizant IRB for determination of whether, how, by whom, and to whom the results should be communicated. Under no circumstance should an investigator performing a noninterventional, retrospective study with archival patient materials attempt to contact patient sources directly.

In contrast to *genetic test,* I accept the argument that *genetic information* should be broadly defined, both to reflect its reality and to enable legislation that prevents discrimination based on genetic information in the insurance and employment markets to be maximally effective in accomplishing that crucial public policy objective. The absence of a consistent definition severely limits the effectiveness of many of the existing state bills and statutes aimed at preventing genetic discrimination; and attempts to customize the definition are almost invariably problematic, as in recent drafts of the Domenici bill (S. 422). A comprehensive and appealing definition was proposed in 1995 by the National Action Plan for Breast Cancer and the ELSI Working Group (Hudson et al. 1995): "'Genetic information' refers to information about genes, gene products or inherited characteristics that may derive from the individual or a family member." The secretary of health and human services has endorsed that definition, and it was incorporated into many of the bills that were introduced in the 105th Congress, for example, H.R. 306 (Slaughter) and H.R. 341 (Stearns).

A point of clarification is necessary here. By endorsing a broad definition of *genetic information* for purposes of antidiscrimination legislation, I acknowledge acceptance of a form of genetic exceptionalism, but for that single purpose only. I make that compromise for two major reasons. First, I strongly believe that with the rapid emergence of genetic and molecular medicine in coming decades and with the absence of comprehensive health insurance coverage for all Americans, to permit access to health care to remain at risk on the basis of probabilistically predictive genetic information is plainly bad health policy and bad social policy, as well. Second, I am persuaded that the fear of discrimination in health insurance coverage and employment serves as a major stimulus of the public's anxiety about permitting ready access to tissue samples and other forms of genetic information.

Redirecting the Deliberations on Informed Consent. In the public deliberations to date about genetic privacy and the use of human tissue samples in research, a tremendous amount of effort and emotion has been expended in trying to deal with the thorny issues of informed consent. I suggest that the effort has been largely misdirected, a position expressed by others (Buchanan 1998; NCVHS 1997), and that continuing to wage battles over the bioethical and legal niceties of informed consent is not an effective or practicable way to resolve the central issue of how to give due respect to the protection of patient privacy while at the same time continuing to facilitate research access to all archival patient materials, irrespective of whether those be tissue samples or medical records. The United States has made a substantial and enduring public commitment to the support of biomedical and related health research, and compelling public interest exists to ensure that any new federal statutes or regulations dealing with medical (or genetic) information privacy do not unreasonably impede continuing access to the materials required to pursue that research effectively.

Current regulations and practices involving informed consent have been elaborated with great care to deal with circumstances involving the direct interaction or intervention of investigators with prospective human research subjects. In that context the stringent requirements of the informed consent protocol are both reasonable and feasible; that is, it is possible to describe to a prospective subject in detail the objectives, duration, procedures, risks, and benefits of the study, whether the investigators have conflicting personal, financial, or other interests, and any other matters deemed appropriate by the investigators or the cognizant IRB. *None* of those circumstances obtains with respect to tissue specimens excised for medical reasons or, for that matter, to any archival patient materials that are used in what is commonly referred to as secondary research. With respect to tissue samples, which can be preserved indefinitely and have been accumulating in the archives of some major academic medical centers for over 100 years, it is impossible to predict or even imagine, let alone describe, the kinds of research questions or experimental technologies with which those samples may be involved in future years. Accordingly, for secondary research on human tissue specimens or other archival patient materials, an informed prospective consent of the stringent kind routinely used in direct interactions with research subjects is quite literally an impossibility—a position that the common rule recognizes in its provisions dealing with archival materials.

To deal with that problem, advocates have proposed three alternative generic approaches to the issue of consent. The first is embodied in the generally, but not universally, accepted proposition that different levels of stringency of informed consent should apply to the use of human tissue samples in research, depending on whether the collections are to be prospective or retrospective and whether the samples are linkable to specific individuals. Most agree that a minimum burden of informed consent—or none—should apply to samples that are "anonymous," that is, completely and irreversibly unlinked to patient sources. Thus, maximum facilitation of tissue-based research might be accomplished if only the tissue samples to be used could be routinely anonymous. That is an unrealistic expectation, however, given that the national tissue archive, which has been and remains the predominant source of human tissue samples for research, overwhelmingly comprises specimens that were removed for medical reasons and must remain identifiable for future access on behalf of the patient, just as the patient's medical record must remain identifiable.

Moreover, although sample anonymity is compatible with some research protocols, all laboratory standardization procedures, and all uses of tissues in professional education, in many studies—especially in pathogenesis research—it is often necessary to obtain additional or follow-up clinical information to establish the diagnostic or prognostic significance of the findings. Indeed, absent the availability of such correlative information, the utility of the primary data would be severely compromised. Thus, although the investigator might only rarely need to know the identity of individual subjects, the research materials, whether they be tissue samples or medical records, must remain linked to their patient sources. Accordingly, the anonymity route does not offer a general solution to the problem of informed consent.

The second generic approach would call for specific reconsent from the patient sources or their next of kin for each research use of a tissue sample or other archival material. Although such a requirement might appear theoretically attractive and comforting to the proposers, it is utterly unfeasible as a general rule. Given the age distribution of the collection of stored specimens typically involved in those studies and considering the large populations that must often be accessed either to amass sufficient experimental material or to ensure that the sample is appropriately structured and unbiased, those experienced in such studies agree that the requirement for serial reconsent would be unduly burdensome, forbiddingly costly, impractical administratively, and susceptible to uncontrolled selection that could totally invalidate the research findings. Exactly the same considerations apply to most epidemiological and health services research protocols that are based on the study of archival patient records. The third generic approach would turn the concept of stringent informed consent on its head. That approach would replace an informed discussion of a specific research protocol with emotionally laden speculations about hypothetical future uses of tissue specimens and invitations to prospective subjects to direct, limit, and prohibit the future uses of their specimens. Such an approach, if widely implemented, could cripple the future research flexibility and utility of the human tissue archive. Moreover, to introduce such speculative discourse into a clinical setting typically freighted with anxiety is to risk transforming a process designed to accomplish informed consent into one that too often would lead to "uninformed denial."[2]

Strengthening Compliance with Human Subjects Regulations and IRB Oversight of Tissue Research. To retain public trust, the research community must strengthen its compliance with human subjects regulations and IRB oversight of research involving human tissue samples. First and foremost, and notwithstanding powerful opposition from all corners of the health care industry, a compelling case can be made for extending the protec-

2. I first heard this expression used in a presentation by Dr. Robert J. Levine, professor of medicine at Yale University, at a meeting of the organization Public Responsibility in Medicine and Research in 1996.

tions of the common rule to all human subjects research conducted in the United States, regardless of site or source of funding. Regrettably, the Clinton administration and the 105th Congress did not make a concerted political effort to accomplish that objective (for example, by backing the Glenn bill).

General agreement appears to be emerging that the current practice of burying in routine hospital admission and operative consent forms broad language that permits the use of medically removed tissue specimens for research and education is inadequate and even deceptive and should be changed. Exactly *how* to change the practice remains very controversial, however, both politically and scientifically and has been the focus of intense debate with respect to the several medical privacy bills introduced in the 105th and 106th sessions of Congress. To provide context for the ensuing discussion of that matter, I summarize the general features that characterize the medical privacy bills, which are highly pertinent to the subject of this chapter.

The bills are generally concerned with strengthening the security of and limiting unauthorized access to individually identifiable medical information and mandating the implementation of fair information practices (Department of Health, Education, and Welfare 1973). Accordingly, the bills require stringent, time-limited, case-specific, and revocable authorization for each instance of disclosure and auditable seven-year record keeping of such disclosures. The bills all require implementing strong physical, administrative, and technical measures to enhance the security and integrity of medical information, and they all mandate some combination of stiff criminal, civil, and administrative penalties for violations. Notwithstanding those provisions, however, and consistent with the recommendations of the secretary of health and human services (Shalala 1997b), all the bills recognize the existence of particular private and public functions that are deemed to be compelling and require relatively unhindered access to medical information, such as emergencies, public health reporting and investigations, reporting of adverse events to the FDA, the needs of medical examiners, the legal system, and the judiciary. All the bills, under threshold conditions that vary somewhat among them, exempt such access from the requirement of specific authorization.

Another set of generally acknowledged "compelling" needs justifying relatively unhindered access includes medical treatment and payment and *health system operations*—a marvelously flexible and capacious term that includes such things as risk assessment and case management, oversight, accreditation, certification, licensing activities, utilization review, financial audits, management-initiated studies of utilization, quality, medical efficacy, and cost-effectiveness. A common method of dealing with that set of needs is to require from each patient at the time of first encounter with a new health plan or provider an upfront general authorization permitting all necessary access without further need of specification or authorization. Although there is general, albeit sometimes reluctant, acceptance of the legitimate need for relatively free access to medical information for those purposes—the details of what is actually subsumed under the term *health system operations,* however, remain controversial—debate continues about the proposed mechanism. Many argue that to require the mandated general authorization as a quid pro quo for access to medical care is coercive and hardly compatible with "informed consent."

An alternative proposition, which is often dubbed "statutory authorization" and is equally controversial, deems that the very act of signing up for access to medical care conveys "constructed consent" for access to medical information without further specific authorization.[3] Despite prolonged and good-faith efforts, a consensus solution that would simultaneously meet the needs of the industry and win broad public favor has not yet emerged from those debates.

With respect to access to medical information for research, the bills generally defer to existing common-rule provisions for activities that fall within the federal regulations; how the bills handle research that falls outside the reach of the common rule is quite variable. Some simply avoid dealing with the issue by calling for further studies by some combination of the General Accounting Office, the NCVHS, or the secretary of health and human services. The Bennett bill (S. 881), in contrast, lays out a detailed scheme of oversight and management of the disclosure of individually identifiable medical information for purposes of research that, while not inviting new federal regulation, nonetheless attempts to conform closely to provisions of the common rule. It is important to recognize that the Bennett language has been crafted with the active participation and support of the involved industries and has been endorsed by them.

3. The House Republican bill, the Patient Confidentiality Assurance Act, which died in the 105th Congress, contained significant provisions dealing with medical information. It explicitly established a "statutory authorization" to permit access—without specific patient authorization—to individually identifiable medical information for purposes of treatment, payment, and "health care operations." The draft also preempted all state laws dealing with the specific medical information provisions contained in the bill. A similar provision of "statutory authorization" is contained in H.R. 448 (Bilirakis), which has been introduced in the 106th Congress.

Finally, it is important to note that all but one of the bills (S. 573, Leahy-Kennedy) insist on treating medical information unitarily: they recognize no subsets of medical information about mental illness, genetic disorders or predispositions, or sexually transmitted diseases for special treatment. In particular, the medical information privacy bills show no deference to genetic exceptionalism. Although all the bills attempt to establish a uniform national legislative framework to deal with medical information privacy and access, they vary markedly in their approach to the thorny issue of preemption. At the extremes, the Bennett bill calls for strong—or what is colloquially referred to as "floor-to-ceiling" preemption of existing state laws dealing with those matters, while the Leahy-Kennedy bill would establish only a floor of federal protections and permit individual states to enact more protective, and doubtlessly restrictive, legislation.

I review those features because it is likely that some form of federal legislation governing the privacy of medical information will be enacted in the 106th Congress. I also wish to make two observations. First, notwithstanding the firm legislative desire to strengthen the security of medical information, there is no avoiding the existence of a myriad of private and public functions that are deemed to have a compelling need for relatively unrestricted access to that information. Second, how to meet those needs remains highly controversial. In particular, the mechanism of a one-time general or a constructed authorization for access, even to meet the essential operational needs of the health care delivery system, has not won universal approbation.

I have earlier (Korn 1996, 1997, 1998a, b) argued that research on medically excised human tissue samples—whether or not the samples must be linkable to their patient sources, that is, putatively "identifiable"—should continue to be governed by an appropriately crafted general consent protocol that provides clear "notice" by means of a general description of the secondary research uses of tissue samples and their correlated medical records and the importance of that usage to continuing medical progress. The consent request language should be clear and simple, based on fact and reason—not speculation and emotion—and structured on the premise that the human tissue archive has always been and should remain a public resource dedicated to the public good, not, like a savings bank, a depository of private property.

It is instructive here to digress briefly to review the issue of tissue ownership, which plays such an important role in the position of genetic privacy advocates and, together with genetic exceptionalism, is the centerpiece of the proposed Genetic Privacy Act (Annas, Glantz, and Roche 1995b). That issue continues to be murky and ill-defined in the absence of a coherent and dispositive body of case law (OTA, 1987; Perley 1992, 337 n. 13). The common law clearly recognizes a person's tissues and organs as property and gives individuals the right to deal with their disposition in such specific instances as donated blood and blood products, organ donations for transplantation, and, more recently, with the advent of assisted reproductive technologies, frozen deposits of sperm, ova, and zygotes. In addition, the Uniform Anatomical Gift Act expressly recognizes an individual's right to determine the manner of disposition of his own body. On the other hand, with respect to tissue samples excised in the course of providing medical care (Bierig 1993), the matter is much more ambiguous. The only pertinent case to date—which addressed the issue of property rights somewhat narrowly as between a "donor" of tissue materials and investigator(s) who, through research, may derive from the tissue cell lines or other products of commercial value—is the celebrated case of *Moore* v. *Regents of the State of California,* which was decided ultimately in 1990 by a divided California Supreme Court; the U.S. Supreme Court denied certiorari.

The *Moore* court majority ruled on the one hand that the defendant researcher had indeed failed to provide the plaintiff "donor" with adequate informed consent about the surgical and medical procedures that were undertaken in the course of providing medical care. On the other, the court held that there was no basis for the claim of "conversion" because the removal of the plaintiff's cells and tissues had effectively extinguished his property interest in those materials, their contents, and any products to be derived from them. In its decision, the California court stated:

> No court . . . has ever in a reported decision imposed conversion liability for the use of human cells in medical research. . . . In effect, what Moore is asking us to do is to impose a tort duty on scientists to investigate the consensual pedigree of each human cell sample used in research. To impose such a duty, which would affect medical research of importance to all of society, implicates policy concerns far removed from the traditional two-party ownership disputes in which the law of conversion arose. *Invoking a tort theory originally used to determine whether the loser or the finder of a horse had the better title, Moore claims ownership of the results of socially important medical research, including the genetic code for chemicals that regulate the functions of every human being's immune system* [emphasis added].

Notwithstanding the strong rhetoric, and although never countermanded, the *Moore* decision has generated controversy and conflicting commentaries (see, for example, Lin 1996), in large part centered on the argument expressed in one of the court's dissenting opinions:

> [T]he [majority] opinion fails to identify any statutory provision or common law authority that indicates that a patient does not generally have the right, before a body part is removed, to choose among the permissible uses to which the part may be put after removal.

In other words, although it may be the fact, as the majority argued, that no prior judicial determinations would favor Moore's claim, none opposed his claim.

Confounding the issue in the *Moore* case was the fact that the litigation centered on the issue of commercial exploitation and the (property) right of the plaintiff to equitable participation in the fruits thereof. Would the claim of conversion have been applicable in the *Moore* case or even raised if the excised tissues had been used exclusively to expand scientific knowledge and not purposefully to develop a source of potentially valuable commercial products? Or suppose that researchers have no prior intention whatever of developing patentable products but, during the course of their studies, such opportunities arise. Would a claim of conversion be applicable in such an instance? To my knowledge, neither the unchallenged *Moore* case nor any other case has authoritatively dealt with that question. At least before the opportunities presented by the advent of biotechnology and tissue-based "genetic research," however, a commonly accepted interpretation (Melton 1997) was that a patient's consent to a medical procedure implied his abandonment of any property interest in the excised specimens.

In the new age of genetic medicine, uncertainty and controversy about the matter of tissue ownership will likely continue until a dispositive body of case law emerges. To be clear, however, I strongly oppose the proposition that patient sources retain property interests in their residual tissue samples. The establishment of such a proposition would undermine the historic status of the human tissue archive as a *public* research resource and have a chilling effect on human tissue–based research and technology transfer. Such an established proposition would likely introduce a golden age for lawyers but, like a form of perverse alchemy, would help transform the golden age of biology into dross.

Instituting a General Authorization Approach for Access to Tissue Samples and Medical Records for Research. I continue to believe that the general authorization approach could be made to work if that were the optimum outcome achievable, and it would certainly be far preferable to any requirement for project-specific, stringent authorization for access to tissue samples and medical records for research. Yet, during my continuing deliberations and legislation jockeying on the privacy of medical information, I have become increasingly drawn to the arguments raised by researchers who study populations—especially epidemiologists and public health and health services researchers—that even a relatively small percentage of nonrandom decliners has the potential over time to skew databases severely. That can make the databases incapable of supporting some studies and can introduce into others severe biases that will distort conclusions, lead to public misinformation, or even invalidate entire studies.

A possibly informative experiment to test that hypothesis is currently underway in Minnesota, which in

1997 implemented a new law requiring (general) authorization for any use of medical records in research. Investigators at the Mayo Clinic—which has produced countless numbers of important retrospective clinical and epidemiological studies over the decades and which founded and has been a major beneficiary of the highly valued, federally funded Rochester Epidemiology Project (Melton 1997)—have expressed concern about the substantial institutional costs of implementing the law and have noted the alarming database skewing that results from the nonrandom nature of the patient cohort that declines authorization (Jacobsen et al. 1998; Melton 1997; Yawn, Jacobsen, and Geir, Jr. 1997). In March 1999 an AAMC colleague and I visited Minnesota to examine the effects of the statute first hand. From meetings with administrators and investigators from the Mayo Clinic, University of Minnesota, and private sector health care delivery systems and health plans, it became clear that the statute, and the way it was being interpreted and implemented in the different institutions, posed a serious threat to medical research and especially to epidemiological, health services, and public health research (Bender 1998; McCarthy et al. 1999; Korn and McCabe 1999).

Accordingly, I have become increasingly supportive of the position that noninterventional research on archival patient materials (records and tissue samples) collected during the course of medical care should be permitted *without authorization* under three circumstances. The first is that a federal privacy law, with its substantial protections and sanctions, is in force. The second is that the involved institutions are in compliance with the requirements of the federal statute and have in place (a) administrative, technical, and physical safeguards to protect the confidentiality, security, accuracy, and integrity of individually identifiable health information; and (b) written policies and procedures to ensure the security and confidentiality of individually identifiable health information and to specify permissible and impermissible uses of such information for medical research. The third circumstance is that an IRB (or comparable independent body) has approved the research proposal and determined that (a) the proposal requires access to individually identified medical information; (b) the proposed research could not reasonably be done without the information; and (c) the researchers have an appropriate plan for protecting the confidentiality of the medical information as well as the research data to be obtained in the course of the study.

That position is substantially in agreement with those expressed in the current policy statements of the American College of Epidemiology[4] and the Association for Health Services Research,[5] and it is also essentially identical to that articulated by the NCVHS (1997) in its recommendations to the secretary of health and human services. In that report the NCVHS stated:

> The Committee recognizes the conflict between research and privacy, but requiring patient consent as a condition of researcher access is impractical and expensive. It would also most likely stop a significant amount of useful investigation. This is not in the health interest of individual patients or the general

4. Personal communication from Robert A. Hiatt, M.D., Ph.D., chair of the Policy Committee at the American College of Epidemiology.

5. Personal communication from Michael Stafford, then acting CEO of the Association for Health Services Research.

> population. Patient privacy interests are adequately protected by independent review of research protocols, the earliest possible removal of identifiers, prohibitions against the use of research records for actions against patients, and strict penalties against researchers who violate the rules.

In interpreting those recommendations, it is important to understand that most of the federal privacy bills that have been introduced to date do not include encrypted medical information that does not directly identify individuals or cannot readily be used to do so, and for which the decryption key is secured, to fall under the requirements of the bill. But, importantly, the bills do declare that unauthorized attempts to use encrypted information to identify individuals are federal crimes. In other words, to the extent that research proposals can be accomplished in secure environments with encrypted tissue samples and medical records, the bills would not require authorization in any event and would not require an IRB or any other independent body to review the proposals. That approach is purposefully intended to provide strong encouragement to investigators to use encrypted research materials to the maximum possible extent.

Privacy advocates and many disease advocacy and patient groups—whose arguments fundamentally rest on the individual right to privacy on the one hand and on fears of insurance and employment discrimination and social stigmatization on the other—will surely continue to oppose vehemently any legislative attempt to permit access without authorization to individually identifiable, archival patient materials for purposes of research. Those groups will oppose unauthorized access even though it would be strictly conditioned on the presence of federal statutory protections and institutional policies and practices of the kinds outlined above. Fears of hurtful reprisals are serious matters, but they would be better addressed explicitly by appropriate legislative remedies and enlightened social policies than by encumbering access to materials essential for medical research of great public benefit. The privacy arguments, like any that are ideological and cloaked in the rhetoric of ethics or morality, are much more difficult to resolve by common methods of political negotiation and compromise. Witness the interminable national travails over the matters of abortion and embryo and fetal research.

But with respect to promoting law and policy that facilitate access to archival patient materials for research, a strong, compelling, ethical argument exists that has not been sufficiently made in the current public debate. Every individual who seeks medical care is the immediate and direct beneficiary of the knowledge that has been created over generations from the retrospective study of medical records and stored tissue samples (Korn 1996, 1998a, b). From that base of knowledge the health care provider obtains his understanding of the nature and behavior of the disease process(es) at hand, the comparative benefits and risks of available therapeutic interventions, and the factors that will shape the prognosis. The knowledge that informs and determines much of the course of contemporary medical care is almost entirely experiential, and it has been accumulated systematically from the "pro bono contributions" of innumerable past patients—to the immeasurable benefit of all those afflicted individuals who will follow them.

Since every individual is a direct beneficiary of the historic medical knowledge base, the ethical principle of distributive justice would suggest that everyone should be obligated to contribute to that base.

Enhancing the Security of Medical Information Created, Used, and Stored in Research. No recommendation that would reduce or eliminate the need for authorization for access to archival genetic or medical information would be appropriate or feasible without the significant strengthening of the protection of all medical information that is contemplated in the pending federal privacy bills. But beyond those protections, an even greater enhancement of the security of medical information created, used, and stored in research is possible and would be of immeasurable benefit in helping to allay public unease, sustain public confidence in medical research, and ensure that research requiring access to archival tissue samples and patient records carries no more than minimal risk. An attractive, effective, and feasible mechanism for accomplishing that objective would be the creation by regulation of a new form of institutional assurance[6] mechanism, or incorporation into a new federal medical information privacy statute of language, modeled on the existing statutory certificate of confidentiality (42 U.S.C. sec. 241(d); Earley and Strong 1995). Under the assurance mechanism, all institutions and entities that engage in human subjects research, including secondary research on archival patient materi-

6. Institutions that receive federal research funds must file with the sponsoring agency annual assurances attesting that the institution is in compliance with federal regulations covering such topics as scientific misconduct, conflict of interest, the use of experimental animals, human subjects, radioactive chemicals in research, and hazardous waste disposal.

als, and that have in place the protective measures described in the preceding section (which may well be mandated imminently by federal law) would be eligible to receive from the federal government—or another appropriate entity in the case of institutions that currently fall outside the common rule and are loath to expand government oversight of their activities—a blanket protection of their entire human subjects research database. The statutory approach would extend that protection to all human subjects research databases, independent of funding sources or venues, without precondition.

Legislative History. The Comprehensive Drug Abuse Prevention and Control Act of 1970 (P.L. 91-513) creating the certificate of confidentiality was enacted as part of the "war against drugs" to facilitate research into drug abuse by Vietnam War combatants and veterans by protecting investigators against compulsory disclosure of the identities of research subjects. In 1974 the protections were broadened to cover mental health research, including research on alcohol and other psychoactive drugs (Comprehensive Alcohol Abuse and Alcoholism Prevention, Treatment, and Rehabilitation Amendments), and in 1988 they were incorporated into the Public Health Service Act and greatly expanded to encompass biomedical, behavioral, clinical, or other research (Health Omnibus Programs Extension, P.L. 100-607, § 163).

Issuance of the certificate, for most of its existence at the discretion of the secretary of health and human services but very recently devolved to the individual institutes of NIH, is to individual applicants and is therefore (in theory) limited to specific research projects. Of particular interest is the fact that federal funding of the research is not a prerequisite for application or award.

Among the categories of information deemed sensitive are "information that if released could reasonably be damaging to an individual's financial standing, employability, or reputation within the community," and "information that normally would be recorded in a patient's medical record, and the disclosure of which could reasonably lead to social stigmatization or discrimination." The cited language could not have been better crafted to capture the fears expressed by those afraid of leakage and misuse of their genetic, or more generally, medical, information. The certificate authorizes researchers

> to protect the privacy of individuals who are the subject of [the] research by withholding from all persons not connected with the conduct of [the] research the names or other identifying characteristics of such individuals. Persons so authorized . . . may not be compelled in any Federal, State, or local civil, criminal, administrative, legislative, or other proceedings to identify such individuals (42 U.S.C. § 241(d)).

The only statutory exceptions are for audits by the funding agency, or by the FDA, but the Department of Health and Human Services additionally ordered in a 1991 policy memorandum (Mason 1991) that the certificate must not preclude public health reporting of communicable diseases required by law.

Commentators (Nelson and Hedrick 1983) have noted an ambiguity in the statutory language as to whether the certificate in fact protects the entire research record or only the "names or other identifying characteristics" of research subjects. Only two challenges are known to have been adjudicated to date, both in state courts in New York. In one case, *People* v. *Newman* (1973), the state's highest court interpreted the scope of protection unambiguously and broadly to include all research records. The second case, *People* v. *Still* (1975), may seem at first approximation to have enunciated a narrower interpretation, but that case is made at best ambiguous—and I believe irrelevant—by the fact that the defendant volunteered that he was participating in a drug research program and thereby was deemed to have waived his protection under the certificate of confidentiality statute. As a consequence, the issue of whether the defendant's medical records could be disclosed was ruled to be controlled by a different statute (42 U.S.C. § 290dd-2), which permits the disclosure of medical records of patients in federally funded drug abuse programs in three situations, including upon a court order if good cause is demonstrated. The statute states:

> In assessing good cause the court shall weigh the public interest against the injury to the patient, to the physician-patient relationship, and to the treatment services (42 U.S.C. § 290dd-2(b)(2)).

It is important to recognize that the statutory language establishing the certificate of confidentiality does *not* contain a "good cause" exception; accordingly, the certificate's protection can be interpreted as absolute, as it was in the case determined by the New York high court. Although the case record bearing on that important issue is scant, legal scholars seem generally to agree that the protections afforded by the certificate do extend to the complete research record. As Nelson and Hedrick (1983) have stated, "[To interpret the certificate of confidentiality to] limit coverage to identifying information alone could make a sham of the statutory protection."

I propose that a mechanism should be established, modeled on the certificate of confidentiality, that would expand the near-absolute protections of the certificate from individual projects to institutions and state clearly that the protections embrace all medical information created, stored, and used in research—that is, the entire human subjects research database. The implementation of such a mechanism would provide important reassurance to human research subjects that directly identified or reasonably identifiable medical information used and developed in research could not be forcibly disclosed—without their express authorization—to anyone, including insurers, employers, family members, health care organizations, or government agencies. Although some continue to question whether such a mechanism would withstand the persistence of a demanding prosecutor or judge, to the best of my knowledge, the absolute protection afforded by the certificate of confidentiality has never been breached during its twenty-nine-year existence,[7] and that is all the more remarkable given the fact that the legislation that required the certificate was originally enacted to protect research subjects who were engaged in patently illegal behaviors.

Building a Fire Wall between Research Data and Patient Care Records. It is noteworthy that the purview of the certificate of confidentiality is explicitly restricted to research records and does not extend to information that is considered part of normal patient care and available

7. The apparent impregnability to date of the certificate of confidentiality was corroborated in a personal communication from Joe S. Cecil, J.D., Ph.D., director of the Scientific Evidence Program, Research Division, and Fazal Khan, legal intern, at the Federal Judicial Center, Washington, D.C.

in the medical record. That raises an important point because the medical record often contains genetic information, whether derived from the medical or family history, from routine clinical laboratory studies, or from the direct analysis of DNA sequences. That is the reason that any discussions of genetic privacy and the protection of genetic information must inevitably confront the medical record and the much broader issues of the privacy, confidentiality, and accessibility of medical information created in the course of the delivery of clinical care. That is also the reason that the population of researchers who would be subjected to any new statutes and regulations purported to be directed narrowly at strengthening the security of genetic information would extend well beyond those who use human tissue samples to include all those investigators—epidemiologists, health services and public health researchers, and others—for whom medical records and databases constitute the primary research materials.

Much of the public pressure for enacting federal legislation governing the privacy of medical information arises from awareness that although a complex web of existing state statutes, regulations, accreditation standards, and professional codes addresses the confidentiality of medical information, medical records have in fact become routinely—and to many, alarmingly—accessible. It is veritably impossible today to obtain medical care from a health care professional or institution without on first encounter having to sign a bland general release authorizing widespread access to one's medical record for insurance and health care purposes. And, as I pointed out earlier in this chapter, none of the pending medical privacy bills, notwithstanding impassioned rhetoric and the best of intentions, has been able effectively to corral the broad access to medical information that will always be required for medical treatment, payment, and health system operations and for public functions deemed to be essential.[8] As a consequence, it is simply impossible even to try to extend to the clinical record the very tight protections afforded research records by the certificate of confidentiality. Put differently, we can achieve a much higher degree of protection for medical research data than for clinical data, and I argue that we should and must strengthen and sustain the public trust required if we are to have continued ready access to the public's medical archives.

Accordingly, I would argue that, at least for the foreseeable future, we must build a fire wall around research data to ensure that they do not freely commingle with patient care records. Although such a fire wall can readily be conceived in specific circumstances to pose troubling consequences and to raise on occasion thorny ethical issues of investigator responsibility, I believe that such instances can and must be managed by carefully crafting research protocols and calibrating the levels of informed consent appropriate to those protocols at the outset or, when necessary, by seeking the guidance of the IRB after the fact. For example, if a research protocol requires direct interaction or intervention with a

8. It is of interest to note that the Health Information Privacy Model Act adopted in 1998 by the National Association of Insurance Commissioners has a broad exception to the requirement of specific authorization for access to individually identifiable medical information for activities dubbed "insurance functions." That exception, in its sweeping breadth, is entirely analogous to the exception for "health system (or plan) operations" that can be found in almost every federal medical information privacy bill introduced into Congress to date. Clearly, in dealing with such a very complex and sensitive issue, each major participant strives mightily to protect its own interests.

human subject or if an archival study anticipates the need to communicate with research subjects for any reason, including the possible communication of research findings, then the full burden of the common rule must apply: there must be specific informed consent and full IRB review. If, on the other hand, a protocol that does not anticipate communication with research subjects leads to findings that the investigators believe are pertinent to the subjects' health and well-being, the investigators should bring the matter to the cognizant IRB for disposition. Under no circumstance should an investigator conducting a noninterventional retrospective study ever attempt to communicate with a research subject directly, nor should research data ever be transmitted directly to a clinical record. In my judgment the alternative—to permit any regular conduit to pass information from the research to the clinical record—would fatally breach the protective net that is the heart of the proposed assurance mechanism and would imperil the ability of an investigator to give assurances of confidentiality and of no more than minimal risk of breaches to prospective research subjects.

In dealing with such a difficult, emotionally laden issue, it is well to maintain perspective. Considering all the necessary handling and transmission of medical information that is—and will inevitably continue to be—required by the day-to-day operations of the health care system, the probability of leakage of sensitive medical information is orders of magnitude greater with clinical data than with research data, whose use and accessibility are far more circumscribed. In fact, all the anecdotal reports of leakage—some of them unacceptably careless, others unconscionably mean and hurtful—that so titillate the press and disproportionately fixate Congress and state legislators have involved clinical records, not research records. Indeed, in its report to the secretary of health and human services, the NCVHS explicitly stated that it had been unable to find documentable instances of breaches of confidentiality resulting from researchers' use of medical records.

Creating a Uniform Federal Standard for Medical Privacy. Because research on archival patient materials often requires the culling of samples from large databases residing in different institutions and often located in many different states, I strongly support the arguments of the NCVHS and the secretary of health and human services and the effort of most pending federal medical privacy bills to preempt the existing hodge-podge of state legislation and create a uniform federal standard. While recognizing the considerable political forces that have already gathered in opposition to strong, "floor-to-ceiling" preemption,[9] the high stakes amply justify the effort. In a recently published policy statement, the Association of American Medical Colleges (AAMC 1997) asserted:

> Vital public purposes are served by the availability of medical information that covers the full spectrum of human experiences with health and disease over time. These purposes

9. Most constituencies support the notion of federal "floor preemption" to set a uniform national threshold of medical information privacy protection. In contrast, some patient and disease advocacy groups and health care providers, who believe that information about specific illnesses can be better protected by state laws, as well as the National Association of Insurance Commissioners and the National Conference of State Legislatures, strongly oppose "floor-to-ceiling preemption."

> include the better understanding of disease causes and processes, health care delivery practices, health care outcomes, health care organization, financing, regulation and accreditation, and the quality and efficiency of health care delivery. These public purposes are sufficiently compelling that any new legislation or proposed regulations must assure the continued availability of and access to medical information for these purposes.

Ensuring Representation of All Stakeholders in the Medical Privacy Debate. Finally, in an earlier publication (Korn 1996), I lamented the absence of an effective coordinating process to guide the public debate and ensure that its resolution strikes the proper balance between the conflicting interests of individual privacy and the public benefits of genetic (and more broadly, medical) research. Until very recently, those issues had been examined largely by a process in which the larger biomedical and health research communities had insufficient participation. At least some of the more egregious flaws, oversights, and cumbersome features of the various proposals put forth almost surely resulted from inadequate knowledge and insufficient informed input into the process. A better process with more adequate representation of stakeholders and fuller solicitation of input from all constituents might have helped to ensure a more inclusive and balanced discussion and framed a debate that too often seemed rudderless and—perhaps in part for that reason—very susceptible to premature political intercession. The process might then have succeeded in setting the parameters in an important area of emerging science policy and in providing a credible basis for whatever statutory, regulatory, or less formal mechanisms ultimately are deemed necessary.

At the time of this writing, the matter of researchers' access to stored human tissue specimens rests in the hands of the National Bioethics Advisory Commission, whose report is now expected in the summer of 1999. The commission's recommendations will have to be integrated in some workable manner with the more general provisions encompassed in pending federal legislation governing the privacy of medical information, which has generally, thus far, steadfastly refused to segregate medical information into different components for purposes of differential regulation or protection. How those orthogonal approaches to enhancing the security of and controlling access to medical information in its myriad of forms play themselves out will have an enormous impact on the future conduct of medical research.

Concluding Thoughts

The history of science is replete with examples of particular topics and observations that have affronted commonplace dogma or religious beliefs, aroused public or ecclesiastic hostility, and led to attempts at censorship. The relatively recent past provides examples of the fact that various factions in society have often deemed the study of the "genetic substance"—as a metaphor for a deeper and broader set of concerns about the nature and sanctity of human life—to be such a problematic topic. Thus, in the mid-1950s, when the existence of a bacterial enzyme that could copy DNA and was thus potentially capable of creating DNA was first described, many lay and religious circles denounced the work as trespass into an area of knowledge not meant for human exploration and understanding.

Less than twenty years later, in the early 1970s, when dramatic scientific advances revealed the basis of recombinant DNA technology, an even louder and more widespread outburst of public concern arose that was fueled by remarkably vivid fantasies of promiscuously unrestrained genetic recombination across species barriers that would have made Hollywood scriptwriters proud. Public meetings were held in many campuses and university communities, and legislative proposals were entertained and even enacted to curtail or prohibit the conduct of such experimentation within particular community boundaries (for example, Cambridge, Massachusetts). The scientific community's self-imposed brief moratorium on such research largely mitigated but did not eliminate that unrest. Following the moratorium, the landmark Asilomar Conference proposed a set of ground rules and procedures for public guidance of the future conduct of experiments that quickly transformed biomedical science and spawned the biotechnology industry. To this date, and notwithstanding the dire predictions of many commentators at the time, I am unaware of a single consequence adverse to either the public health or the environment that can be attributed to the discovery and vigorous exploitation of recombinant DNA science and technology.

Now, the increasing power and expanding boundaries of scientific exploration have once again come into conflict with strong, deeply held public convictions. The current disquiet traces its pedigree back to the 1950s and is reflected in the rhetoric and recommendations of the scholars and working groups that have been studying the use of human tissues in research. The roots of the current unease are certainly understandable. At a practical level, the inappropriate use of genetic information can have dire consequences for individuals and families with respect to such critical matters as employment, insurance coverage, and reproductive decisions. At a much deeper level, some people perceive even obtaining and studying such information as the ultimate assault on the fundamental right of individuals and their families to privacy and to a sense of personal equanimity about themselves and their life plans. Thus, the issues under debate are very serious, and the stakes are very high. On one side stands the protection of the sanctity of the self from what is feared to be a new wave of scientific infringement without license. On the other stand unparalleled opportunities to illuminate the fundamental nature of human diseases and offer new hope for therapeutic and preventive strategies that could help to mitigate some of the major scourges of mankind.

The rapid pace of scientific and technological advances, especially in the biomedical and information sciences, and the continuing transformation of the health care delivery system make it inevitable that challenging issues of genetic privacy, informed consent, and the ownership and custodianship of patient data will continue to arise. Society will continue to be vexed by troubling questions that lie at the boundary between its commitment and respect for individual autonomy and privacy and its compelling interest in promoting the benefits that flow from the generous public investment in research. It is important that the policies that emerge from the current debate be attentive to patient privacy and respectful of informed consent. But at the same time, those policies must not unduly encumber access to the invaluable research treasure represented by the vast archive of human tissue specimens and medical records. Moreover, those policies must not impede the continuing accession and accessibility of that archive for the support of future research. Policies must thoughtfully and sensitively balance the competing values of private interest and public benefit and not be unduly or precipitously shaped by emotionally charged and often exaggerated public fears.

References

American College of Medical Genetics (ACMG), Storage of Genetics Materials Committee. "Statement on Storage and Use of Genetic Materials." *American Journal of Human Genetics* 57 (1995): 1499–1500.

American Society of Human Genetics (ASHG). "ASHG Report: Statement on Informed Consent for Genetic Research." *American Journal of Human Genetics* 59 (1996): 471–74.

Americans with Disabilities Act, U.S.C. §§ 12101–12213 (1994).

Annas, George J., Leonard H. Glantz, and Patricia A. Roche. "Drafting the Genetic Privacy Act: Science, Policy, and Practical Considerations." *Journal of Law, Medicine & Ethics* 23 (1995a): 360–66.

———. *The Genetic Privacy Act and Commentary.* Boston: Health Law Department, Boston University School of Public Health, 1995b.

Association of American Medical Colleges (AAMC). "Patient Privacy and the Use of Archival Patient Material and Information in Research." Memorandum No. 97-21, April 4, 1997.

Bender, Alan P. "Public Health Protection *vs.* Informed Consent." *Minnesota Medicine* 81 (May 1998): Commentary.

Bierig, Jack R. "Who Owns Human Tissue Removed in Surgery or at Autopsy?" Presentation made at the American Society of Clinical Pathologists/College of American Pathologists Fall Meeting, Orlando, Fla., 1993.

Buchanan, Allen. "An Ethical Framework for Biological Samples Policy." Paper commissioned by the National Bioethics Advisory Commission, 1998.

Clayton, Ellen W. "Prospective Uses of DNA Samples for Research." In *DNA Sampling: Human Genetic Research—Ethical, Legal, and Policy Aspects,* edited by Bartha Maria Knoppers. Boston: Kluwer Law International, 1997.

Clayton, Ellen Wright, Karen K. Steinberg, Muin J. Khoury, Elizabeth Thomson, Lori Andrews, Mary Jo Ellis Kahn, Loretta M. Kopelman, and Joan O. Weiss. "Informed Consent for Genetic Research on Stored Tissue Samples." *Journal of the American Medical Association* 274 (December 13, 1995): 1786–92.

Comprehensive Alcohol Abuse and Alcoholism Prevention, Treatment, and Rehabilitation Act Amendments of 1974, Pub. L. No. 93-282, 88 Stat. 125 (1974).

Comprehensive Drug Abuse Prevention and Control Act of 1970, Pub. L. No. 91-513, 84 Stat. 1236 (1970).

Council of Europe, Committee of Ministers. "Recommendations of the Committee of Ministers to Member States on the Protection of Medical Data." Recommendation No. R(97)5 (February 1997).

de Gorgey, Andrea. "The Advent of DNA Databanks: Implications for Information Privacy." *American Journal of Law & Medicine* 16 (1990): 381–98.

Department of Health and Human Services. *Final Report of the Task Force on the Privacy of Private Sector Health*

Records. Washington, D.C.: Department of Health and Human Services, September 1995.

Department of Health, Education, and Welfare, Secretary's Advisory Committee on Automated Personal Data Systems. *Records, Computers, and the Rights of Citizens*. Washington, D.C.: Government Printing Office, 1973.

Earley, Charles L., and Louise C. Strong. "Certificates of Confidentiality: A Valuable Tool for Protecting Genetic Data." *American Journal of Human Genetics* 57 (September 1995): 727–31.

Equal Employment Opportunity Commission (EEOC). *EEOC Compliance Manual*, vol. 2, section 902, Order 915.002 (1995): 902–45.

European Parliament and the Council of Europe. "Directive 95/46EC of the European Parliament and the Council." *Official Journal of the European Communities* No. L281 (1995): 31–50.

Federal Policy for the Protection of Human Subjects: Notices and Rules. 56 *Federal Register* 28,003–28,032 (1991).

Gostin, Lawrence O. "Genetic Privacy." *Journal of Law, Medicine & Ethics* 23 (Winter 1995a): 320–30.

———. "Health Care Information and the Protection of Personal Privacy: Ethical and Legal Considerations." *Annals of Internal Medicine* 127, no. 8, pt. 2 (October 15, 1997): 683–90.

———. "Health Information Privacy." *Cornell Law Review* 80 (1995b): 451–528.

Gostin, Lawrence O., Zita Lazzarini, Verla Neslund, and Michael T. Osterholm. "The Public Health Information Infrastructure: A National Review of the Law on

Health Information Privacy." *Journal of the American Medical Association* 275 (June 24, 1996): 1921–27.

Health Insurance Portability and Accountability Act of 1996, Pub. L. No. 104-191, 110 Stat. 1936 (1996) (codified as amended in scattered sections of 18, 26, 29, and 42 U.S.C.).

Hudson, Kathy L., Karen H. Rothenberg, Lori B. Andrews, Mary Jo Ellis Kahn, and Francis S. Collins. "Genetic Discrimination and Health Insurance: An Urgent Need for Reform." *Science* 270 (October 20, 1995): 391–93.

Institute of Medicine, Committee on Regional Health Data Networks, Division of Health Care Services. *Health Data in the Information Age: Use, Disclosure, and Privacy.* Edited by Molla S. Donaldson and Kathleen N. Lohr. Washington, D.C.: National Academy Press, 1997.

Jacobsen, S. J., Z. Xia, M. E. Campion, et al. "Potential Impact of Authorization Bias on Outcomes Research." Manuscript submitted for publication, 1998.

Korn, David. "Contribution of the Human Tissue Archive to the Advancement of Medical Knowledge and the Public Health." Paper prepared for the National Bioethics Advisory Commission, January 1998a.

———. "Dangerous Intersections." *Issues in Science and Technology* (Fall 1996): 55–62.

———. "Genetic Privacy and the Use of Archival Patient Materials in Research." (April 1998b). Available at <http://ASIP.uthsca.edu/korn.html>.

———. "Genetic Privacy and the Use of Archival Patient Materials and Data in Research." In *Proceedings of the Public Health Conference on Records and Statistics and the Data Users Conference on Partnerships, Technologies and*

Communities: Evolving Roles for Health Data. CD-ROM No. 1, July 1997.

Korn, David, and Colleen McCabe. "Confidentiality of Medical Records. AAMC Report on the Minnesota Experience." (May 1999). Available at <http://www.aamc.org/advocacy/issues/research/minreport.htm>.

Knox, E. G. "Confidential Medical Records and Epidemiological Research: Wrongheaded European Directive on the Way." *British Medical Journal* 304 (6829) (March 21, 1992): 727–28.

Lin, Michael M. J. "Conferring a Federal Property Right in Genetic Material: Stepping into the Future with the Genetic Privacy Act." *American Journal of Law & Medicine* 22 (1) (1996): 109–34.

Lowrance, William W. *Privacy and Health Research: A Report to the Secretary of Health and Human Services.* May 1997. Available at <http://aspe.os.dhhs.gov/datacncl/PHR.htm>.

Lynge, E. "European Directive on Confidential Data: A Threat to Epidemiology." *British Medical Journal* 308 (1994): 490.

Mason, James O. "Certificates of Confidentiality—Disease Reporting." Policy memorandum, August 9, 1991.

McCarthy, Douglas B., Deborah Shatin, Carol R. Drinkard, John H. Kleinman, and Jacqueline S. Gardner. "Medical Records and Privacy: Empirical Effects of Legislation." *HSR: Health Services Research* 34 (April 1999, Part 2): 417–25.

Melton, Joseph L., III. "The Threat to Medical-Records Research." *New England Journal of Medicine* 337 (November 13, 1997): 1466–70.

Moore v. *Regents of the University of California,* 793 P.2d 479 (Cal. 1990), *cert. denied,* 499 U.S. 936 (1991).

Murray, Thomas H. "Genetic Exceptionalism and Future Diaries: Is Genetic Information Different from Other Medical Information?" Pp. 60–73 in *Genetic Secrets: Protecting Privacy and Confidentiality in the Genetic Era,* edited by Mark A. Rothstein. New Haven: Yale University Press, 1997.

National Association of Insurance Commissioners (NAIC). "Health Information Privacy Model Act." Draft, July 2, 1998. Available at <http://www.naic.org/committee/modelaws/models.htm>.

National Commission for the Protection of Human Subjects of Biomedical and Behavioral Research. *The Belmont Report: Ethical Principles and Guidelines for the Protection of Human Subjects of Research.* DHEW Publication No. (OS)78-0012. 44 *Federal Register* 23,192–23,197 (1997).

National Committee on Vital and Health Statistics (NCVHS). "Health Privacy and Confidentiality Recommendations of the National Committee on Vital and Health Statistics." Approved on June 25, 1997. Available at <http://www.mssm.edu/medicine/medical-informatics/ncvhs.html>.

National Institutes of Health (NIH), NIH-DOE Working Group on Ethical, Legal, and Social Implications of Human Genome Research. *Genetic Information and Health Insurance: Report of the Task Force on Genetic Information and Insurance.* May 10, 1993.

National Research Council, Committee on Maintaining Privacy and Security in Health Care, Applications of the National Information Infrastructure, Computer

Science and Telecommunications Board, Commission on Physical Sciences, Mathematics, and Applications. *For the Record: Protecting Electronic Health Information.* Washington, D.C.: National Academy Press, 1997.

Nelkin, Dorothy, and M. Susan Lindee. *The DNA Mystique: The Gene as a Cultural Icon.* New York: W. H. Freeman, 1995.

Nelson, Robert L., and Terry E. Hedrick. "The Statutory Protection of Confidential Research Data: Synthesis and Evaluation." Pp. 213–36 in *Solutions to Ethical and Legal Problems in Social Research,* edited by Robert F. Boruch and Joe S. Cecil. New York: Academic Press, 1983.

Office for Protection from Research Risks (OPRR). "Human Genetic Research." Pp. 5-42 to 5-63 in *Protecting Human Research Subjects: Institutional Review Board Guidebook.* Washington, D.C.: NIH Office of Extramural Research, 1993.

Office of Technology Assessment (OTA). *New Developments in Biotechnology: Ownership of Human Tissues and Cells—Special Report.* OTA-BA-337. Washington, D.C.: Government Printing Office, 1987.

———. *Protecting Privacy in Computerized Medical Information.* TCT-576. Washington, D.C.: Government Printing Office, 1993.

Patient Access to Responsible Care Act of 1997 (PARCA), H.R. 1415, 105th Cong., 1st Sess. (1997a).

Patient Access to Responsible Care Act of 1997 (PARCA), S. 644, 105th Cong., 1st Sess. (1997b).

Patients' Bill of Rights Act of 1998, H.R. 3605, 105th Cong., 2d Sess. (1998a).

Patients' Bill of Rights Act of 1998, S. 1890, 105th Cong., 2d Sess. (1998b).

People v. *Newman*, 298 N.E.2d 651 (N.Y.), *cert. denied*, 414 U.S. 1163 (1973).

People v. *Still*, 389 N.Y.S.2d 759 (1975).

Perley, Sharon Nan. "From Control over One's Body to Control over One's Body Parts: Extending the Doctrine of Informed Consent." *N.Y.U. Law Review* 67 (1992): 335–65.

"Pharmacoepidemiologists Moving to Protect Access to Medical Record Information." *Epidemiology Monitor* 18 (1997): 1–3.

Powers, Madison. "Privacy and the Control of Genetic Information." Pp. 77–100 in *The Genetics Frontier: Ethics, Law, and Policy*, edited by Mark S. Frankel and Albert H. Teich. Washington, D.C.: American Association for the Advancement of Science Press, 1994.

Public Health Service Act, 42 U.S.C. § 241d (1991).

Reilly, Philip R. "Panel Comment: The Impact of the Genetic Privacy Act on Medicine." *Journal of Law, Medicine & Ethics* 23 (Winter 1995): 378–81.

Reilly, Philip R., Mark F. Boshar, and Steven H. Holtzman. "Ethical Issues in Genetic Research: Disclosure and Informed Consent." *Nature Genetics* 15 (1997): 16–20.

Presidential Commission on Consumer Protection and Quality in the Health Care Industry. *Report of the Advisory Commission on Consumer Protection and Quality in the Health Care Industry*. Washington, D.C.: Government Printing Office, 1998.

Rothenberg, Karen H. "Genetic Information and Health Insurance: State Legislative Approaches." *Journal of Law, Medicine & Ethics* 23 (Winter 1995): 312–19.

Rothenberg, Karen, Barbara Fuller, Mark Rothstein, Troy Duster, Mary Jo Ellis Kahn, Rita Cunningham, Beth

Fine, Kathy Hudson, Mary-Claire King, Patricia Murphy, Gary Swergold, and Francis Collins. "Genetic Information and the Workplace: Legislative Approaches and Policy Challenges." *Science* 275 (March 21, 1997): 1755–57.

Shalala, Donna. *Confidentiality of Individually Identifiable Health Information: Recommendations of the Secretary of Health and Human Services Pursuant to Section 264 of the Health Insurance Portability and Accountability Act of 1996.* September 11, 1997b. Available at <http://aspc.os.dhhs.gov/adminsimp/pvrrec0.htm>.

———. *Report of the Secretary to the President: Health Insurance in the Age of Genetics.* Washington, D.C.: Department of Health and Human Services, July 1997a.

Task Force on Genetic Testing. *Promoting Safe and Effective Genetic Testing in the United States: Final Report of the Task Force on Genetic Testing.* Edited by Neil A. Holtzman and Michael S. Watson. NIH-DOE Working Group on Ethical, Legal, and Social Implications of Human Genome Research (ELSI Program), September 1997b. Available at <http://www.nhgri.nih.gov/ELSI/TFGT_final/>.

———. "Protection of Human Subjects: Suggested Revisions to the Institutional Review Board Expedited Review List." 62 *Federal Register* 60,604–60,611 (1997a).

Vandenbroucke, J. P. "Maintaining Privacy and the Health of the Public." *British Medical Journal* 316 (1998): 1331–32.

Westin, Alan F. *Computers, Health Records, and Citizen Rights.* National Bureau of Standards Monograph 157. Washington, D.C.: Government Printing Office, 1976.

World Medical Association. "Declaration of Helsinki: Recommendations Guiding Medical Doctors in Biomedical Research Involving Human Subjects, 1989 Revision." Available at <http://www.ncgr.org/gpi/odyssey/privacy/HelDec.html>.

Yawn B., S. J. Jacobsen, and G. R. Geir, Jr. "Who Provides and Who Denies Authorization for Medical Record Review for Medical Research." In *Proceedings of the Public Health Conference on Records and Statistics and the Data Users Conference on Partnerships, Technologies, and Communities: Evolving Roles for Health Data.* CD-ROM No.1, July 1997.

THREE

The Social Implications of the Use of Stored Tissue Samples: Context, Control, and Community

Karen Rothenberg

David Korn sets up some presumptions, assumptions, and paradigms that need to be challenged. The debate, concern, and angst surrounding genetic privacy forces us to ask, Are patient groups, researchers, and privacy advocates all talking a different language? Are we all approaching privacy from different perspectives? How can we get to a point where we can really listen to what others are saying?

My husband always tells me that I worry too much about the privacy of genetic information and the uses to which it is put, and that most people walking down the street do not think about the issues at all. I got a reality check when my mother said, "I read in the newspaper about a cancer gene—the so-called Jewish cancer gene for colon cancer. A few months ago, I read about another gene for breast and ovarian cancer. Why," she asked, "are all those studies looking only at the Jews?"

I explained to her that genetic researchers were studying many different gene pools and ethnic groups, such as the Finnish and the Amish. I also told her that researchers are particularly interested in populations characterized by "founder's effect."[1]

She said, "I really don't understand any of that. I see in this newspaper article that the researchers used some samples from people who had undergone testing for Tay-Sachs syndrome. Didn't you get a Tay-Sachs test before you were pregnant with Andrea?" When I said yes, she asked, "Did they use your sample?"

I said, "I have no idea. The samples are anonymized." She asked, "How could they use anonymized samples?"

I thought this was a good reality check. Most of us have been in the hospital, or we have had a relative in the hospital. The first time I heard that my tissue sample was being used for research, when I had not given it to anybody for that purpose, I questioned the propriety of that. Would I have agreed to submit a tissue sample if I had known that it was going to be used in research? Probably, although I am more cautious now than I was a few years ago. In fact, the little empirical data we have indicates that most of us, if asked, would consent to allow our tissue samples to be used in research.

The key question for us is, What if we are not asked? What implications does that have for a public backlash against science? How much of the public's trust and support for science is going to be further eroded if consent procedures are not clear? Patients should know that when

---

1. The genetic composition of any particular population is known as its gene pool. When a few individuals leave a large population and start their own new isolated population, a gene that is relatively rare in the large gene pool can become common in the new gene pool. Founder's effect occurs when a particular gene becomes disproportionately represented in a population because many of its members can trace their lineage back to a few individuals.

they have a blood or tissue sample taken in the course of a clinical visit, a portion of it is stored for their benefit. After all, tissue samples are stored primarily for the patient's protection, for accreditation, and for liability reasons.

Research is a public benefit that Dr. Korn, and indeed many of us, want to support. But how do we communicate that value without creating a public backlash? The first question we have to examine then becomes, Why do we even care about genetic privacy? Some of us are afraid that the information will be misused. That concern causes us to commingle concerns about genetic discrimination and genetic privacy. It is hard, as a public policy matter, to separate genetic discrimination and privacy because an individual who believes that he has been discriminated against has the difficult task of proving it; a person may not even know that his information was actually used. Therefore, it becomes important to protect access to the information, not just to erect safeguards against discrimination. We create a false dichotomy when we separate privacy issues from discrimination protections; the two have to be taken together as a package.

A second reason that we care about genetic privacy is that it encourages research. If we cannot assure patients that their genetic information will be kept private or at least that physicians will do the best they can to keep it confidential, patients will not want to be research subjects, at least in situations in which we require informed consent. If that indeed is the rationale for why we care about genetic privacy, then why have Dr. Korn and many other scholars set up a paradigm that balances concern for privacy with public benefit and research? I do not like that paradigm. It does not encourage broad-based community support for biomedical research. Instead, I would like to see a paradigm that actively builds up public trust so that patients can support our research agenda. We should make the public a partner in biomedical research, rather than create a dichotomy between private rights and public benefit and then try to maintain a precarious balance.

There are three *C*s to think about when discussing genetic privacy. The first is context. The most important reason we care about genetic privacy is that it is a social value. Independent of caring about misuse and independent of caring about our agenda for research, it is just something we, as a community, want to encourage and promote. Genetic privacy is a value that is the basis of being an individual. It is a value that we care about in the context of our family, and it is a value that we care about for our community. Therefore, context matters. It matters whether samples are collected in a clinical context or in a research context. There is a lot of graying around the edges, particularly with genetic testing, but we need to ask, first, what the expectation of the individual is at the time his tissue sample is collected—whether in the clinical context or the research context. How do we recognize what the patient's expectation is and deal with the patient's knowledge needs?

The second *C* is control. If we want patients to support the research establishment, we have to give patients some sense of control. That is what my mother's phone call was about: Who gave the researchers permission to use the Tay-Sachs test results? Did the subjects consent to the use of their tissue samples?

The third *C* is a sense of community. Supporting research in a partnership is a communitarian effort, but that does not mean only recognizing the benefits to the community. It also requires recognizing that there are risks to a community. That was my mother's final point, when she asked why they were picking on us. Nobody is "picking on us." We need to put the "picking on us" in context with the science and the needs of our community.

We should think about the themes of context, control, and community when we reexamine the paradigm that requires us to balance concern for private rights with public benefit and research.

FOUR

The Regulation of Genetic Testing

Michael S. Watson

Medical genetics, including genetic testing, has been among the most rapidly developing areas of medicine. Until relatively recently, it has involved individuals and families with relatively rare disorders. Although in aggregate those diseases affect a large number of individuals, they are individually rare such that there has not been the need for rapid expansion of genetics training programs. We are now moving into an era in which the genetic contributions to common diseases are being identified, and significant proportions of the population may be involved in genetic testing. That testing involves patients in almost every specialty of medicine and requires medical geneticists to be broadly based in many fields. Similarly, such testing requires that knowledge of genetics be distributed throughout medical specialties and primary-care providers. As such, the underlying premise is that medical genetics must be integrated into the broader health care delivery system rather than separated from it by genetics-specific regulations.

This chapter presents a brief overview of the recent history of medical genetics and defines genetic testing. Using the genetic testing for cystic fibrosis as an example, I discuss the highly iterative nature of genetic test development and evolution into clinical use. In tracking the time course of cystic fibrosis testing breakthroughs, I highlight the critical steps. Only with an understanding of the issues underlying those critical steps will appropriate oversight develop.

Medical Genetics in the Health Care System

Recognition of medical genetics was attained in 1991 with the acceptance by the American Board of Medical Specialties of the American Board of Medical Genetics as the certification body for each of the specialties of the field: clinical genetics, clinical cytogenetics, clinical molecular genetics, and clinical biochemical genetics. Medical genetics became the twenty-fourth primary specialty of medicine. Shortly thereafter, and consistent with the organization of other specialties, the American College of Medical Genetics was formed as the arm of the field responsible for standards and practices. Unlike many other specialties, however, medical genetic services have been provided in an interdisciplinary way. That interdisciplinary provision of services has complicated the development of self-regulatory programs because medical geneticists have had a limited impact on those providing services from other specialty areas.

Additional factors have converged with the evolution of this new field to complicate the development of new genetic tests and their transition into clinical service. Not surprisingly, the significant resources invested in the Human Genome Project have lead to the rapid identification of genes causing or contributing to human disease. Those discoveries have come at a pace far beyond that of the training of medical geneticists and genetic counselors. Similarly, because new tests have been developed by using laboratory-based technologies and reagents rather than manufactured test kits, the role of the regulatory system has been minimized, and its development with regard to genetic testing has been slowed. An underlying premise in that regard is that oversight must avoid compromising research and investigation in deference to controlling the practice of medicine. Other important factors that have complicated the transfer of genetic technologies into practice—although not discussed in this chapter—are issues of privacy of medical information, including genetic information, and the use of genetic information to discriminate unfairly against individuals and families.

Concurrent with the evolution of this new field, we have seen health care move from a fee-for-service delivery system controlled by direct providers to a more laissez-faire system in which the health care service and manufacturing industries control service delivery and patients play a more active role in determining their health care choices and options for managing disease. Underlying that transition has been the recognition that there must be some control over the health care budget to contain exploding costs. As we now move into the "genetic age" of medicine, we must determine how we shall provide assurance to the public that the provision of that testing, as well as many other types, will be safe and effective and of clinical utility and that equitable access to genetic medicine is maintained. To a large extent, "problems" in genetic testing parallel those seen in the oversight of medicine in general, while others are specific. How we resolve those concerns will determine the future growth of the field.

What Is Genetic Testing?

Because genetics affects all areas of medicine, a definition of *genetic testing* must be similarly broad but must distinguish aspects of "genetic testing" that raise concern. Because the genetic component of any disease may be a strong or weak contributor to disease development, we should not presume that geneticists should be primarily involved in the delivery of all health care. The definition of *genetic testing* agreed on by the NIH-DOE Task Force on Genetic Testing attempts to acknowledge that fact (Task Force on Genetic Testing 1997):

> Genetic test—The analysis of human DNA, RNA, chromosomes, proteins, and certain metabolites in order to detect heritable disease-related genotypes, mutations, phenotypes, or karyotypes for clinical purposes. Such purposes include predicting disease, identifying carriers, establishing prenatal and clinical diagnosis or prognosis, and monitoring, as well as carrier, prenatal and newborn screening, but they exclude tests conducted purely for research. Tests for metabolites are covered only when they are undertaken with high probability that an excess or deficiency of the metabolite indicates the presence of heritable mutations in single genes.

The definition of *genetic testing* is critical because it has legislative, regulatory, and other implications. When defined excessively broadly, every medical record in the United States contains "genetic information." When defined overly narrowly, many tests for heritable traits fall through the cracks. Hence, the goal of the definition is to distinguish clearly between genetic tests for germ-line, heritable changes and those tests for acquired changes causing disease, such as infectious agents or the somatic changes characteristic of cancers.

A genetic predisposition to cancer can be identified through a genetic test and would be within that definition, while a genetic test for mutations that diagnose or stage a particular cancer would not be. Complicating that definition is the fact that the same test may do both and is distinguished by the tissue that is used as the substrate for the test. Similarly, research and forensic applications of genetic testing are excluded. Of particular importance, however, is the recognition that a genetic test is more than just the technical components of the test. Rather, important preanalytic—for example, phenotypes and ethnicity—and postanalytic—result interpretation—aspects of genetic tests are critical components of a genetic testing service.

A final important distinction that must be made before considering oversight of genetic testing is the inherent transition through which any developing test must move. The initial identification of a gene involved in a genetic disorder is usually in a research stage in which no patient identifiers exist: the gene-disease relationship is established on a population basis. In the second stage the magnitude of the relationship between the gene mutations and the disease are established. That relationship can range from mutations that are deterministic (that is, those with the mutation have or get the disease in question) through those with high penetrance (nearly all with the mutation get the disease) to those gene mutations that have a strong likelihood of leading to disease (for example, BRCA1 testing for inherited breast and ovarian cancer) and to those gene mutations that have relatively minor roles in disease development (for example, cholesterol testing). Once the clinical and analytical performance characteristics of the test are established, the test moves into an area of accepted clinical testing.

The Life of a Genetic Test

For informed decisionmaking about the points in test development that require oversight, it is useful to review the development and transition into service of a genetic test. In 1985 cystic fibrosis was linked to markers on chromosome 7 that allowed the chromosome carrying the mutation to be tracked through families (Tsui et al. 1985). Cumbersome Southern blotting methods and highly complex linkage analysis restricted test provision to a small number of clinical and research genetics laboratories usually directed by individuals certified by the American Board of Medical Genetics. Testing was targeted at families with living affected individuals that had to segregate informative linkage markers. About 80 percent of families were informative, and test accuracy was a function of distance of markers from the cystic fibrosis locus. Over the next few years, there was rapid technological development as nucleic acid amplification techniques were introduced. By the 1990s, numerous methods were available for isolating nucleic acids, amplifying specific sequences, or detecting specific mutations or for all three purposes. Simultaneous with the development and introduction into diagnostics of DNA sequencing methods were a number of methods allowing a gene to be scanned for mutations not preidentified within a family. By 1993, 330 mutations had been described (Zielenski and Tsui 1995) and nearly 30 could be specifically tested clinically. Testing was beginning to move from family-based testing to screening of parents in prenatal settings. In 1997 the National Institutes of Health Cystic Fibrosis Consensus Conference (NIH 1997) endorsed screening in some populations and presented comprehensive data on the differences in clinical sensitivity and specificity in various ethnic groups.

More than 500 CFTR (cystic fibrosis transmembrane regulatory protein) mutations had been described, and testing for as many as 60 mutations was available. We are now on the verge of using chip-based sequencing or hybridization techniques that will allow for the rapid and accurate detection of any mutations described.

Where in the process would oversight be appropriate? The first critical step is the scientific validation of the relationship between the test and the disease. That information became rapidly available to geneticists through peer-reviewed literature and meeting presentations. Although not necessarily synonymous with clinical utility, laboratory diagnosis by linkage in families with affected individuals was almost immediately a possibility, and many took advantage of testing. Experts in the field continue to improve existing testing systems while seeking the mutations involved. Testing was generally offered in an investigational context with informed consent. With the identification of the CFTR gene (Riordan et al. 1989), diagnostic testing became an accepted clinical test when individuals carried one of the known mutations. But investigational testing continued in parallel with diagnostic testing for those without previously described disease-causing mutations. Numerous mutation-detecting systems became available, many with advantages in detecting particular types of mutations. The lack of a true diagnostic testing "gold standard" meant that any tests had to be correlated with a clinical diagnosis—an often inaccurate process. But testing was clearly an important tool in establishing diagnoses before development of full-blown cystic fibrosis and in identifying carriers at risk of having additional affected offspring. The specific tests used were commonly compared to prior test methods such as linkage, even though those predicate tests were inherently less accurate than evolving methods. Systems capable of detecting specific mutations made it possible to move from family-based testing to the testing of individuals. In the absence of known disease-causing mutations, however, investigational tests were still pursued. Over the next few years, the critical questions involved assessing the clinical utility of testing among individuals with differing clinical presentations to identify those settings (for example, prenatal testing, carrier detection in different groups distinguished by ethnicity, reproductive status, or other features, specific and general population screening, and newborn screening) in which testing would be of maximal clinical utility. Even with only six common mutations being tested, testing was accurate and clinically useful to some. Unlike many types of clinical tests, however, the critical questions involved the intended use of the test and the individual populations to which it would be directed.

Should regulation be targeted at laboratory methods, probes for mutations, amplification primers, or reasons for testing? The shift of our health care delivery system from one without many financial constraints to one with significant need for cost-benefit justification suggests that many of the issues we are currently debating involve issues generic to medicine (for example, outcome analysis and cost, privacy of medical information, and unfair discrimination based on health status). Others more specific to genetic testing are those of the personnel requirements for those providing services, the standards that must be followed in analytically validating genetic tests, and the knowledge base of the health care community. In that regard, it is important to remember that many of the concerns that have been expressed about genetic testing stem from the preanalytic and postanalytic parts of the tests (such as confidentiality, accurate result interpretation, and communication).

Regulating Genetic Testing and Health Care

Given that regulation in medicine is intended to ensure that the services provided to the public are safe and effective, it is difficult to argue that government should not have a significant role. That role can include legislation, government structures (for example, the Food and Drug Administration), and the imposition of self-regulation. The level at which such regulation is applied is of critical importance, however. That can occur anywhere from a broad policy level down to the microdetails of provision of a particular test. The goal of linking regulation to the issues of concern in genetics requires that we separate those issues that are generic to medicine from those specific to genetics. Further, we must assess the mechanisms chosen by regulatory agencies to provide oversight. For genetics to be incorporated into the general practice of medicine where its central role in disease and health can be maximally utilized, not all solutions will be regulatory. In fact, significant improvements in the use of genetic testing will come from education of the public and health care providers and the maximal acquisition of data from providers of genetic testing services and researchers to guide our decisions about appropriate, safe, and effective use.

The FDA has traditionally regulated the manufacturers of devices and kits used in medical testing. No precedent exists for the FDA's regulating laboratories offering testing, although the agency's legislative mandate may include such a requirement. The lack of readily available reagents and kit-based technologies for genetic testing combined with the numerous analytical test systems capable of identifying the same genetic changes and rapid improvements in those systems has left the FDA on the sidelines. It will likely be years until the FDA can acquire appropriate staff and develop internal policies to guide it in regulating that field. It is important that those new technologies and tests not be withheld from the public and that further research not be stifled while government plays catch-up. As commonly occurs during such periods of change, professional organizations representing those practicing in the area begin to develop the standards of practice without which informed regulation is unlikely. As testing becomes standardized and kits become available to allow the dissemination of testing away from professional medical geneticists, the FDA's role will expand in keeping with its legislated regulatory mandate.

Many have argued that the way laboratories construct a test system from reagents purchased or made inhouse is within the purview of laboratory practice. Practices are generally regulated through the Health Care Finance Administration by way of the Clinical Laboratory Improvement Amendments (1988). Those amendments include requirements for personnel, quality control and assurance, analytical test validation, and proficiency testing. Recently updated to address concerns about PAP smear analysis, the amendments include little that is specific to genetic testing. Except for clinical cytogenetics, no specific requirements exist within genetic testing. Those are deferred to generic requirements for highly complex areas of testing, in recognition of the extensive training and education required of those practicing in the area.

Most gaps in the regulation of genetic testing occur because the FDA is organized around the assessment of devices and the Clinical Laboratory Improvement

Amendments relate to the methods used in specific areas of testing. Genetic testing involves reagents and devices made in-house and uses generic methods of molecular biology, biochemistry, immunology, and cytogenetics that are applicable to multiple specialties of medicine and can be used for many varying purposes. Thus, current models of regulation of genetic testing are unlikely to target the critical issues. Care must be taken that one use of a genetic test is not held hostage to other potential uses that may be less well resolved. Hence, new paradigms of oversight may be necessary to ensure safe and effective delivery of genetic testing within the system.

When considering the regulation of anything, it is important to keep in mind that regulation has both negative and positive aspects. Though the goal may be safety and effectiveness, it is important to remember that regulation is an industry in and of itself and that it imposes significant costs in time and fees that must be passed on to consumers. While the regulated may perpetuate regulation to reduce competition, the regulator may perpetuate regulation for job security. In the end, regulation must be appropriate and directed at ensuring safety and effectiveness so that testing be of value to individuals and families. At the present time, such fundamental questions as "What is a rare disease?" are addressed with much disparity between regulatory bodies, with prevalence often defined as the number of individuals in the population with the disease. Novel features of genetics and inheritance will have to be considered before such definitions can be reasonably applied to genetic testing.

The Task Force on Genetic Testing

In 1994, the NIH and Department of Energy Working Group on Ethical, Legal, and Social Implications (ELSI) of Human Genome Research convened the Task Force on Genetic Testing to review genetic testing in the United States and to make recommendations to ensure the development of safe and effective genetic tests. The task force defined *safety* and *effectiveness* to include the validity and utility of genetic tests, their delivery in laboratories of quality, and their appropriate use in the health care system. We have already mentioned the definition of *genetic testing* under which the task force operated and have overviewed the potential areas of concern in the development and evolution of a genetic test. Clearly, significant amounts of genetic testing were already being done, and quality assurance and control programs were in place for some testing already in use and others in development. The task force was particularly concerned about new tests for common diseases and susceptibilities of otherwise normal individuals to those diseases for which no predicate tests existed and, therefore, for which no confirmatory tests were available. While many of the task force findings apply broadly to genetic testing, many are driven by concerns about predictive testing.

The task force subdivided its work into three major areas: scientific validity of tests, laboratory quality, and the delivery of the service to consumers. Special attention was paid to rare diseases. The task force made a number of recommendations in each of those areas. Particular overarching principles included the need for informed consent in particular situations (for example, predictive testing or investigational testing), the need to maintain confidentiality of genetic information, and the need to avoid unfair discrimination.

In the area of scientific validity, the task force recommended that the relationship between the disease and the gene and its mutations be established, that tests be validated for each intended use, and that a mechanism to acquire data and information on validity be in place before marketing. Recognizing that data on predictive tests will improve over time with increasing numbers of individuals tested and followed, the task force emphasized the need for organized comprehensive data collection. Instead of regulating use in the investigational stage, which is defined as the stage during which many of the parameters of clinical sensitivity and specificity are established, it will be important to maximize the acquisition of data from that stage to reach decisions rapidly about the relative clinical utility of the tests.

In the area of laboratory quality, many felt that the fundamental regulatory structures were in place. Discussions often involved whether the gaps in the system should be filled by regulation or by developing standards of practice or both. Of particular concern to the task force were the need to improve the preanalytic and postanalytic components of test delivery to ensure the accurate and appropriate communication of test utility to patients and the need for similarly accurate communication of test results and possible interventions. The task force felt that the most direct route to improving those aspects of testing was through significant provider education to ensure that those requiring services are recognized, that proper tests are ordered, that counseling and informed consent are obtained where needed, and that results are accurately transmitted with changes in the presentation of information from laboratories to providers. Because so many of the concerns in genetic testing relate to the laboratory practices, the task force recommended that the Clinical Laboratory Improvement Advisory Committee consider creating a genetics specialty with specific requirements for training and education in human genetics for genetics laboratory directors and technical supervisors and that proficiency testing be mandated. Given that the FDA will assume an increasing role in assessing genetic test kits and devices, the task force encouraged the FDA to begin to develop internal structures and staff to prepare for the enormous amount of technology and testing on the horizon and similarly to begin to consider methods of triage to ensure that appropriate regulation is applied to the areas of concern.

With respect to the competence of providers, the task force recognized that before 1987 more than 50 percent of the medical schools in the United States offered little human and medical genetics, so that most currently practicing physicians have limited exposure to that complex area. The task force recommended the development of genetics curricula at four levels, including a base level for all physicians and an intermediate level for physicians in specialties that serve patients with problems with significant genetic components—for example, cancer and clinical oncology. In addition, the public's level of education would be enhanced. The fourth level, that of the board-certified medical geneticist, already has well-developed curricula. In response, the American Medical Association and the American Nursing Association have sponsored the National Coalition for Health Professional Education in Genetics.

Within those recommendations the task force encouraged the existing regulatory bodies to assess their roles in oversight and to develop specific standards to guide regulation. Recognizing that many genetic tests raise few concerns and others great concerns, the task force recommended that aspects of genetic tests that indicate the greatest need for careful scrutiny (for example, predictive testing, unvalidated population or newborn screening, lack of confirmatory testing, absence of useful intervention, and significant disease morbidity and mortality) be determined and applied to decision-making about the need for regulatory oversight. Among the most useful activities to guide oversight would be one that the Clinical Laboratory Improvement Amendments have mandated but not fulfilled: an assessment of the impact of proficiency testing, personnel standards, and quality assurance programs on the accuracy of testing. That should be complemented with documentation of the extent to which treatment problems are caused by inaccurate testing and the effect of errors on the testing process.

Many of the recommendations of the task force require coordination among agencies and monitoring of progress toward the task force's goals as opposed to the formation of new bodies. That fact combined with the rapid pace of development of the field led the task force to recommend that there be an advisory committee to the Department of Health and Human Services to provide interagency coordination and to harmonize state, federal, and interstate regulation and policy in genetic testing. The advisory committee should ensure that appropriate bodies (regulatory, professional, consumer, and others) consider licensing, standards and guidelines for practice, information collection, and evaluation and dissemination of results (to ensure accountability) and should assist in the coordination of records storage. The task force recommended that the committee play an active role in monitoring future technologies and practices. Given the breadth and impact of those recommendations, any such advisory body should be broadly representative of the stakeholders in genetic testing.

The task force acknowledged the important role of the professional organizations involved in delivering genetic services to the public. Only with the continued development of standards of practice will appropriate oversight take place. In an area that is so broad and is used for diseases of so many kinds, however, practitioners cannot rely solely on self-regulation. While self-regulation is important, it tends to be narrowly focused on one facet of a test or on technology such as delivery, medical, scientific, research, or other aspects of the service. Effective oversight will likely require a balance between regulators and practitioners. Those recommendations combined with significant educational programs and active data acquisition can move us toward the safe use of genetic tests of proven value without denying access to those services.

References

National Institutes of Health (NIH). *Genetic Testing for Cystic Fibrosis.* NIH Consensus Conference Statement, NIH Consensus Development Conference on Genetic Testing for Cystic Fibrosis, April 14–16, 1997.

Riordan, John R., Johanna M. Rommens, Bat-sheva Kerem, Noa Alon, Richard Rozmahel, Zbyszko Grzelczak, Julian Zielenski, Si Lok, Natasa Plavsic, Jia-Ling Chou, Mitchell L. Drumm, Michael C. Iannuzzi, Francis S. Collins, and Lap-Chee Tsui. "Identification of the Cystic Fibrosis Gene: Cloning and Characterization of Complementary DNA." *Science* 245 (September 8, 1989): 1066–73.

Task Force on Genetic Testing. *Promoting Safe and Effective Genetic Testing in the United States: Final Report of the Task Force on Genetic Testing.* Edited by Neil A. Holtzman and Michael S. Watson. NIH-DOE Working Group on Ethical, Legal, and Social Implications of Human Genome Research (ELSI Program), September 1997.

Available at <http://www.nhgri.nih.gov/ELSI/TFGT_final/>.

Tsui, Lap-Chee, Manuel Buchwald, David Barker, Jeffrey C. Braman, Robert Knowlton, James W. Schumm, Hans Eiberg, Jan Mohr, Dara Kennedy, Natasa Plavsic, Martha Zsiga, Danuta Markiewicz, Gita Akots, Valerie Brown, Cynthia Helms, Thomas Gravius, Carol Parker, Kenneth Rediker, and Helen Donis-Keller. "Cystic Fibrosis Locus Defined by a Genetically Linked Polymorphic DNA Marker." *Science* 230 (November 29, 1985): 1054–57.

Zielinski, Julian, and Lap-Chee Tsui. "Cystic Fibrosis: Genotypic and Phenotypic Variations." *Annual Review of Genetics* 29 (1995): 777–807.

FIVE

Genetic Discrimination

Philip R. Reilly

To assess whether genetic discrimination exists in the United States, I make the following argument. First, little evidence supports the widespread fear that people who undergo genetic tests to determine whether they are at increased risk for developing a serious disorder face a significant risk of genetic discrimination. Nevertheless, our society has widespread fear that genetic testing carries with it a genuine risk of discrimination. Indeed, the way in which the problem of genetic discrimination has been portrayed has misled the public about the risks and has perhaps dissuaded some of those who might most benefit from taking the tests from doing so. Furthermore, the laws that states enacted to protect individuals from the threat of genetic discrimination have until recently provided little protection to those who might need it if the threat became real. Recent efforts at the federal level have, however, sharply reduced the risk of genetic discrimination in regard to access to health care and employment.

Because the published articles claiming to show that genetic discrimination is a significant social problem are key to my argument, I assess those reports and review the evidence that the public fears being denied insurance or employment because of genetic test results.

My concern is that a well-intentioned effort to combat a relatively small problem—genetic discrimination—has "demonized" genetic testing and may create a net loss to the health of the American people by turning them away from testing technologies that could help to improve health and sometimes save lives. Those who have emphasized the threat of genetic discrimination—a concern raised repeatedly since the late 1980s—should acknowledge that their fears have not materialized and that, since we have made significant progress in enacting preventive state and federal legislation, those are unlikely to be justified.

The Definition

As with all arguments, much depends on the meaning one gives to words. In discussing *genetic discrimination,* I use the definition offered in a seminal article written by Paul Billings and his colleagues (Billings et al. 1992). They defined *genetic discrimination* as

> discrimination against an individual or against members of that individual's family *solely* because of real or perceived differences from the normal genome of that individual. Genetic discrimination is distinguished from discrimination based on disabilities caused by altered genes by excluding, from the former category, those instances of discrimination against an individual who at the time of the discriminatory act was affected by the genetic disease. (Emphasis added.)

That is, genetic discrimination is discrimination against persons in good health whom a test indicates may become ill (or in some cases become the parent of a child with a genetic disorder).

As far as I have been able to determine, the term *genetic discrimination* first appeared in print in 1985. In that year Dr. Peter Rowley and I organized a workshop on the topic as one of the proceedings for the annual meeting of the American Society of Human Genetics. We used the term then to encompass virtually the same set of issues as would Billings et al. (1992).

I trace the origins of the survey research that Billings and his colleagues undertook to a small meeting in about 1987 of the Genetic Screening Study Group, a Boston-based organization of concerned scientists and others to which I was invited to make a presentation and at which I first met Dr. Billings. At that meeting I suggested that there was a need for a study like the one that they later planned, organized, and conducted (and which they may well have already conceived). In making the suggestion I was motivated by the same concern that motivates me today. No published evidence of genetic discrimination existed, and I wondered whether my worries about the potential for abuse of genetic data were justified. A decade later, I still can find little firm evidence to suggest that genetic discrimination constitutes a sufficiently serious threat to curb an individual's interest in being tested.

Published Evidence of Genetic Discrimination

What is the available literature on genetic discrimination, and what inferences may we fairly draw from the literature? Although there were earlier "thought" papers, efforts to gather data on the dimensions of genetic discrimination began in the early 1990s.

Let us begin with the Billings (1992) study, which I regard as the seminal one. To identify cases of possible genetic discrimination, Billings and his colleagues mailed a solicitation to 1,119 professionals working in genetics and related areas. Similar appeals were printed in the newsletters of several associations for persons with Friedrich ataxia, Charcot-Marie-Tooth disease, and muscular dystrophy. Over a period of seven months the group received forty-two responses, of which they excluded thirteen for failing to meet their definition of *genetic discrimination.* The twenty-nine usable responses reported forty-one incidents of discrimination, of which thirty-two involved insurance and seven involved employment. Of special importance to this discussion, the researchers found that "[p]roblems with insurance companies arose when individuals altered existing policies because of relocations or changes of employers. New, renewed, or upgraded policies were frequently unobtainable even if individuals labeled with genetic conditions were "asymptomatic." The authors concluded that "[t]hese incidents suggest that there are very important nonmedical reasons why individuals may wish to avoid genetic testing." They also argued that existing laws were "inadequate to prevent some forms of genetic discrimination, particularly that due to the health insurance industry" (Billings et al. 1992). I return to their findings later.

Probably because—unlike studies that offered only speculative discussion of risks—the study by Billings et al. offered real-life stories of genetic discrimination, the study had an important impact. Paul Billings, in particular, became sought after by the media and appeared on such powerful television programs as *Nightline.* Media interest in genetic testing had been growing steadily during the late 1980s and early 1990s, but much of the reporting had focused on the potential for beneficial clinical advance. Work like that of Billings and his colleagues and those who followed in their footsteps helped to elicit more journalistic interest in the risks of genetic discrimination.

In sharp contrast to those who seized the Billings study as alerting the public to a new threat, I maintain that the study is remarkable for how few incidents of genetic discrimination it was able to discover. Requests to over 1,000 professionals working in the field (individuals who talk with scores to hundreds of families at special risk for genetic discrimination each year) generated virtually no responses. Most of the few cases were provided by families of or individuals with a genetic disorder. Since most complaints focused on denial of insurance applications, it was impossible for the researchers to confirm that the respondents accurately described the reasoning of the insurers in question.

The fact that the Billings study was able to report only seven cases of alleged employment discrimination is consistent with a 1983 report from the now extinct Office of Technology Assessment. The report (commissioned by Congress) was based on a survey of the nation's 500 largest industrial companies. Of 366 respondents, only 6 percent said that they had tested or were conducting any genetic tests. Of even greater interest, only 16 percent expressed any interest in testing in the future (Johnson 1991). As I discuss later, a 1997 survey shows that interest in predictive genetic testing among employers and health insurers remains low.

About the same time as Billings was conducting his research, I was investigating genetic discrimination from a different perspective. In 1992 my colleague, Jean McEwen, and I studied all proposed legislation and existing state and federal laws to ascertain the level of interest in and quality of proposals to regulate the use and potential misuse of genetic information. From the large number of bills being introduced and the few newly enacted laws, we concluded that the legislative interest in genetic discrimination had surged. In general, we thought that most state genetic discrimination laws and bills were inadequately drafted to deal with the problems they sought to address. At most, they regulated the use of genetic information by insurers that provided group health policies. Generally, the laws and bills did not reach the large number of self-insuring employers who could circumvent state rules by claiming exemption under the federal Employee Retirement and Income Security Act (ERISA) (McEwen and Reilly 1992). A particularly important issue, then and now, is that a large percentage of the laws lacked clear definitions of terms like *genetic disorder.* It was often difficult to comprehend what problems the laws could solve. Thus, on one important point we agreed with Billings—if genetic discrimination is an emerging social problem, we needed better statutes to deal with it.

Professor McEwen and I also reasoned that, since genetic discrimination was thought to arise mainly in the context of insurance applications, we might get some sense of the dimensions of the problem by surveying life insurance companies about their use of genetic information. We prepared a detailed survey and sent it to a large number of North American life insurance companies. Responses from the medical directors of twenty-seven of them led us to four conclusions. First, few insurers were performing genetic tests, but most were interested in knowing already existing genetic information about applicants. Second, interest among insurers in using genetic tests in the future would likely depend on the amount of coverage sought and the quality of the tests available. Third, little actuarial data existed to support underwriting decisions about specific genetic problems. Finally, a fair amount of subjectivity existed in the decisions of medical directors about applications from persons with uncommon genetic conditions (McEwen, McCarty, and Reilly 1993). On the basis of the medical directors' performance on some test questions, it also appeared that some directors were not well informed about some basic principles of medical genetics.

At about the same time, we also surveyed the insurance commissioners of the fifty states and the District of Columbia to discover whether they had received any complaints of genetic discrimination from consumers applying for life insurance. We received responses from forty-two jurisdictions. Our results indicated that those who regulate the life insurance industry did not perceive genetic testing to pose a significant problem in how insurers rate applicants. In addition, they thought that insurers had the right to request genetic tests. Only a minuscule number of consumers had formally complained to commissioners about the life insurers' use of genetic data (McEwen, McCarty, and Reilly 1992). I also investigated whether any consumers had sued insurers alleging discriminatory practices in underwriting on the basis of genetic information. I found not a single case. Although I regularly seek information about suits, I still am unaware that any have been filed.

During the 1990s, as concern for the possible misuse of genetic information has mounted, the main focus has remained on health insurance. A lack of evidence to suggest that life insurers make much use of genetic information and a general sense that the need for access to health insurance is much more important seem to explain that narrow focus. Most Americans probably think that access to health care should be an entitlement. Although surveys indicate that a majority of the American public also thinks that access to a modest level of life insurance should be universally available regardless of an individual's health status, those same surveys indicate that many fewer Americans are willing to pay even a modest increase in premiums to make that possible compared with the numbers willing to help provide universal access to health care coverage (American Council on Life Insurance 1995).

Concern about the impact of genetic information on access to health insurance and employment does not appear to be as significant in other nations as in our own, but concern abroad is growing. For example, a 1995 survey of consumers in Helsinki found that the most commonly expressed reason for having reservations about genetic testing was that test results might lead to discrimination in employment or insurance (Hietala et al. 1995). Concern is sufficiently high in England that the Association of British Insurers recently took the position that "[a]ny genetic tests which are to the detriment of applicants for life insurance up to a total of £100,000 will be ignored when the insurance is linked to a new mortgage for a home" (Association of British Insurers 1997.)

The report by Paul Billings and his colleagues (1992) has generated much commentary but little subsequent research. What else has been published? Late in 1995 the influential journal, *Science,* published a policy forum entitled "Genetic Discrimination and Health Insurance: An Urgent Need for Reform" (Hudson et al. 1995), which was coauthored by Francis Collins, director of the then–National Center for Human Genome Research. The core of the article's argument was that a "*potential* for discrimination in health insurance coverage for an ever increasing number of Americans" (emphasis added) exists. As evidence, it cited the experience with sickle cell testing in the 1970s, the Billings study, and a then-unpublished work that the authors said had found that among people with a known genetic condition in their family, 22 percent had indicated that they had been refused health insurance coverage because of their genetic status, whether they were sick or not. The authors also recounted two cases of discrimination. Among other broad comments, they stated, "As genetic research enhances the ability to predict individuals' future risk of diseases, many Americans may become uninsurable on the basis of genetic information." They closed by calling for laws that would prohibit essentially any use of genetic information by health insurers.

In 1996 a group comprising some of the same persons who had participated in Billings's 1992 research published a follow-up study entitled "Individual, Family, and Societal Dimensions of Genetic Discrimination: A Case Study Analysis" (Geller et al. 1996). They sent 27,700 survey instruments to "individuals at-risk to develop a genetic condition and parents of children with specific genetic conditions." Of 917 respondents, 455 asserted that they had experienced genetic discrimination and 437 that they had not. After evaluating their responses, the researchers contacted 206 persons from a group of 385 respondents intimately involved with one of four genetic disorders—Huntington's disease, hemochromatosis, mucopolysaccharidosis, and phenylketonuria. Roughly two-thirds were from the Huntington's disease group, which renders that study largely one of a cohort at risk for a single disorder. From among their telephonic interviews, the researchers selected and summarized eighteen cases as examples of the range of genetic discrimination.

Much as in the Billings study, the authors took the respondents at their word. They did not interview the insurers or others about whom complaints were made. Thus, it is impossible to verify the substance of the allegations. But even if we assume that the self-reporting would stand up under close scrutiny, I am impressed that such a large survey of a population especially likely to experience genetic discrimination gleaned such a low response. One can fairly assume that those who believe that they have been discriminated against would be likely to respond to a survey inquiring about such an experience. In that context, it is of note that only about 1.7 percent (455 of 27,700) of those surveyed asserted that they had been harmed.

Some months after publishing the policy forum on genetic discrimination, *Science* published the research to which that piece had alluded, an article entitled "Genetic Discrimination: Perspectives of Consumers" (Lapham, Kozma, and Weiss 1996). The work was based on an analysis of structured telephone interviews with adult volunteers recruited via letters sent to the directors of 101 genetic support groups representing about 585,000 people. Those leaders in turn spread the call for participation. No attempt was made at random sampling from among that larger cohort. A group of 483 persons responded; of those, 336 persons returned the required consent forms and 332 were interviewed (306 by telephone and 26 by written questionnaire). The respondents were primarily highly educated, married, white women with two or more persons in their families who were affected with a genetic disorder.

The term *genetic discrimination* was not used in the interview. The article defined the term to include

> prejudicial actions as perceived by the respondents that resulted from insurers' or employers' knowledge of an individual's genetic condition, carrier status, or presumed carrier status, based on observation, family history,

> genetic testing, or other means of gathering genetic information.

That is, the authors *greatly broadened* the definition from the one used in the Billings study. In a footnote they acknowledged:

> Because the questions on discrimination ask about all family members at once, the questions do not distinguish among: (i) the direct consequences of ongoing genetic disease or conditions, (ii) the effect of genetic disease on other family members, and (iii) the consequences of genetic information gained through testing.

Perhaps the core finding in the study, certainly the one that the media seized on, was that 22 percent of those interviewed claimed that they or a member of their family had been refused health insurance as a result of a genetic condition in the family. On the other hand, 76 percent responded that they had not been refused health insurance for that reason. The authors did not attempt to cross-check a sample of the responses. Thus, the 22 percent figure is an upper bound that would almost surely drop significantly if further investigations were made. Because of the broad nature of the questions, the article permits few inferences. It does little more than establish that, just as is the case with disorders such as cancer and heart disease, people who have serious genetic disorders may experience difficulty in obtaining various kinds of insurance. The study also shows a widespread fear that insurers will use genetic information to deny coverage to persons with or at increased risk for serious disorders. Although the authors acknowledged the limits of the study, they may be faulted for using the term *genetic discrimination* in their title when they were departing from the definition then in use. From a methodological perspective, I am surprised that the study passed peer review for *Science.*

What do those few studies tell us? One view, apparently the dominant one, if one scans the legislative horizon, is that they raise enough evidence of the existence of genetic discrimination to support prompt legal action to combat discrimination before it grows further. A second view is that the survey research was designed and carried out in a way to yield results from which there is no basis on which to generalize. A third view is that the studies document evidence of practices that are routine and legal within our insurance system, but that some, perhaps many, think reflect serious flaws in it. That is, the researchers found evidence that in some instances underwriting decisions have been made on the basis of genetic information. The studies contain no evidence to suggest that those actions constituted illegal practices by insurers. A fourth possible view is that the research, despite a design that strongly favored the discovery of cases of genetic discrimination, is remarkable for how little evidence of such practices it discovered. Not enough information exists to know which of those views most closely approaches reality, but I favor the last.

The Public Perception

For some time a substantial fraction of the American public has thought that a serious risk of genetic discrimination associated with genetic testing exists, and evidence abounds that the percentage of those who have that concern is on the rise, largely because of the avalanche of attention the media give to genetic discrimination. Over the past five years I have received scores of calls from the national television networks, local TV and radio stations, and print journalists about the topic. Time and again the main question to me has been, "Can you put me in touch with a person who has suffered from genetic discrimination?" What emerges from those conversations is that the journalists are desperate to find, but have great difficulty in finding, citizens who will make that claim.

My assertion about public fears should not go unsupported. A review of relevant national public opinion polls provides that support. Between 1986 and 1994 at least five large national surveys of the American public dealt with issues of genetic testing: a 1986 Harris poll, a 1991 Gallup/Columbia University poll, a 1992 National Opinion Research Center poll, a 1992 Harris/March of Dimes poll, and a 1993 Harris/Westin poll. Most of them sought to ascertain the level of public knowledge about genetics and genetic testing. The 1986 poll asked few questions relevant to *genetic discrimination,* a term not then in wide use, but the poll did find that 66 percent of those asked thought that "genetic engineering" would improve their lives. The 1991 poll found that 66 percent of respondents thought that genetic screening would on balance do more good than harm. The 1992 Harris poll discussed privacy issues such as sharing sensitive genetic information within extended families and government access to genetic information and asked, "Do you believe that new types of genetic testing should be stopped until these types of privacy issues are settled, or not?" Nearly four in ten (38 percent) of those asked said that new tests should be halted pending resolution of privacy concerns. More religious, less educated, and older Americans were most likely to favor such a freeze (Westin 1997). I take that response as a strong indicator of a high level of concern about the dangers of genetic discrimination.

Perhaps more informative is the 1992 National Opinion Research Center survey that focused on genetic information in the workplace. Ninety percent of those asked rejected "the position that employers should have the right to screen prospective employees for genetic defects before hiring them"; 60 percent did so even when the issue was posed in terms of possible health risks to the worker; 75 percent of the respondents favored giving workers exclusive control over access to records. The survey results led the National Opinion Research Center to conclude that "attitudes toward genetic screening in the workplace . . . are overwhelmingly negative" (Westin 1997).

The Harris/Westin poll conducted in the spring of 1993 asked questions about employment and health insurance. Pollers conducted 1,000 telephone interviews of a sample of employees and managers in private-sector employment and 300 interviews with corporate human resource executives and found that all three groups strongly (91 percent) opposed the use of genetic information in hiring to reduce health benefit costs. They also strongly (86 percent) opposed genetic testing by health insurance companies to assist them in making underwriting decisions. When the genetic test was presented as a way to ascertain special risks to employees in the workplace, however, a majority of both human resource executives (56 percent) and employees/managers (63 percent) favored testing. A similar majority favored the proposition that employers should be required "to provide other, safe jobs at comparable pay to employees at special risk." In that survey 85 percent of employees/managers and 88 percent of human resources officials did not believe that an employer should be allowed "to refuse to hire someone who has a genetic condition that predicts the likelihood of contracting a major disease late in life." Eighty-eight percent of employees/managers and 93 percent of human resources executives opposed the use of genetic blood tests to determine the likelihood of developing inherited diseases or major disabilities (Westin 1997). The results of that survey are heartening. The very people that consumers suspect are likely to misuse genetic information in the workplace—the bosses—are by an overwhelming majority saying that such information should not be used to discriminate in employment.

National opinion polls seek trends in thinking and ask broadly based questions. The available survey data indicate that in the early to mid-1990s the American public strongly opposed the use of genetic testing in hiring decisions and in decisions about enrolling people in health care plans. Little evidence indicates what their views were concerning the proper role of genetic testing in life insurance. The fact that almost four of ten respondents were willing to halt the use of new genetic tests until privacy issues were settled is perhaps the strongest indicator of the level of concern about—broadly speaking—privacy issues.

The most recent national opinion survey, conducted in 1997 on behalf of the National Center for Genome Resources in Santa Fe, confirms that the American people still have a high level of concern about misuse of genetic information, but it also indicates that the vast majority of employers and health insurers do not think that they should be permitted to use genetic information to discriminate. Unless one is willing to argue that the respondents are not answering honestly, the results of the National Center for Genome Resources' survey (1997) should strongly reassure those who fear genetic discrimination.

State Genetic Discrimination Laws

During the 1990s state legislative interest in regulating the use of genetic information grew dramatically, primarily in regard to its use in deciding access to health care coverage. In the past few years many articles have been published that trace the evolution of those laws and summarize their content (Rothenberg 1995), so I do not repeat that effort here. Suffice it to say that a substantial minority of states (about twenty) now have statutes intended to limit the use of genetic information by health insurers or to give individuals control (by requiring consent for transfer of information) of the use of genetic data compiled about them. Some statutes do both. The laws enacted before 1995 are, generally speaking, of limited scope and offer limited protections. The laws enacted since 1995, perhaps owing to the influence of a model genetic privacy act published by George Annas and colleagues in February of that year (Annas, Glantz, and Roche 1995), are significantly more comprehensive and generally provide more authority to the individual whose tissue is being tested. The state-based legislative efforts to combat the risk of genetic discrimination are less important than recent developments at the federal level. That is mainly so because most of them were written to limit the use of data by health insurers, not to give individuals a general authority to control the flow of genetic data generated about them.

Those laws are new, and it is not yet possible to measure their effects. The best one can do is to ask what impact they are likely to have. To answer that question one must determine to whom and what the laws apply and what uses of genetic information they actually restrict. Thus far, those laws generally apply only to health insurers' use of genetic data. I discern no strong trend to enlarge the scope of those bills to regulate life, disability, or other forms of insurance. Furthermore, the existing genetic discrimination laws are limited to regulating the use of genetic test results. Generally speaking, they do not prohibit insurers from inferring genetic information from medical records or from asking questions about family history (Reilly 1998).

Those laws are irrelevant to Americans who are covered under Medicare and largely irrelevant to those covered under Medicaid. They are also of little importance to the roughly 65 percent of Americans who obtain health care coverage for themselves and their families through employer-based group plans that do not rate the health status of individuals. For otherwise healthy persons in whom a genetic test has suggested an increased risk of developing a disease in the future, the state laws are valuable in dissuading plan managers from expanding the reach of "preexisting condition" clauses. But for many of those persons who are at risk, the onset of illness is likely to occur, if at all, long after preexisting condition exclusions have been tolled, so there was little incentive to the insurer to so label them. More important, a recently enacted federal law, the Health Insurance Portability and Accountability Act of 1996, takes control of that issue by forbidding group health insurers from using genetic information for that purpose.

Ironically, the new state genetic discrimination laws may be most relevant to a fraction of the population that is unlikely to need them. Some of the statutes cover persons seeking to purchase individually underwritten health insurance. That class, which numbers less than 10 percent of Americans, is steadily declining, largely because of the high cost of purchasing such policies. In general, those who can still afford that option tend to have the resources to obtain the coverage they desire despite negative genetic information. The statutes will make that task a bit easier.

Taking into account the population of the states in which laws have been enacted and estimating the fraction of the population that has health care coverage to which the laws are relevant, I calculate that those laws protect significantly less than 10 percent of Americans.

Since 1995 some of the state bills have become more expansive in scope. In addition to regulating or prohibiting the use of genetic information by health insurers, a few have proposed to curtail or forbid employers or life insurers or both from seeking genetic information. One approach (incorporated in New York and Oregon law and proposed in several different bills, including, notably, New Jersey, where it was stopped only by gubernatorial veto) has been to define a tissue sample, the genetic information derived from it, or both as the "property" of the individual from whom it was taken. That challenges (if not overturns) a century-old custom in medicine that discarded blood and tissue may be reasonably used for teaching and research purposes. It is, of course, correct, that those who own property can sell it or give it to others. Nevertheless, the public policy stance that tissue or information derived therefrom is property raises concerns in the academic and commercial research communities about how to secure intellectual property rights to work done with such tissue. To secure their property rights, will researchers have to redesign consent forms so that they read more like licensing agreements? To those who scoff that the worry is misplaced, I can reply with confidence that it is, nevertheless, real.

To me the most worrisome feature of some state and federal bills has been the possibility that, if enacted, they would have a chilling effect on genetic research.

That has not yet happened, but the pharmaceutical industry spearheaded the drive that at the last minute persuaded Governor Whitman to veto the New Jersey bill that had adopted the property approach. Some pharmaceutical companies are studying state laws to decide whether the firms should continue to conduct certain kinds of research in particular jurisdictions. Those firms worry that the cost of complying with genetic discrimination laws is sufficiently high that it might be easier just to avoid a local problem. That has also been an issue in Oregon and Massachusetts. One recent draft of a bill introduced in the U.S. Senate by Pete Domenici (R-New Mexico) in 1997 contains provisions to regulate the conduct of human genetic research that have worried many of the human geneticists who have read them (the Genetic Privacy and Nondiscrimination Act of 1997).

Federal Efforts to Prevent Genetic Discrimination

Two important developments at the federal level—if taken together—will significantly reduce the (small) threat of genetic discrimination in health insurance and employment. The first is the Equal Employment Opportunity Commission's March 15, 1995, decision to issue a guideline (Leary 1995) interpreting the reach of the Americans with Disabilities Act of 1990 in regard to people who have been found to carry a gene that puts them at increased risk—or that others perceive to put them at risk—for a serious disorder. The second federal development is the enactment of the Health Insurance Portability and Accountability Act of 1996 (HIPAA).

In a nutshell, the EEOC guideline asserts that a person who thinks that he was discriminated against in hiring or promotion on the basis of genetic information may claim the protections of the ADA. Such an expansive reading of the ADA has not been tested in court. Given that there is virtually no legislative history to support the guideline and that the ADA is under fire for other reasons, I am not convinced that the guideline could sustain judicial review. But, if it did, it would be a truly powerful disincentive to discriminate.

HIPAA represents an important legislative advance in the effort to diminish the threat of genetic discrimination in health insurance. Designed to maintain the integrity of health insurance coverage for workers and their families as they migrate from one job to another, the act includes a provision that forbids the use of a genetic condition as a "preexisting condition"—a category that triggers a waiting period for coverage. That provision of the act addresses head-on the very concern that Billings and his colleagues stressed as the most significant problem raised by genetic discrimination in health insurance (Billings et al. 1992). It is curious that many people, including consumer groups, genetic counselors, and clinical geneticists, have expressed grave doubts about the protections conferred by HIPAA. True, the act does not extend protection to those without health insurance, and it does not control the costs of being in the system. But, given the fact that an ever-growing percentage of Americans have group coverage and given the firm manner in which the act closes off concern about ERISA, that is a helpful development.

Congress may not be ready to pursue a global solution to the threat of genetic discrimination, but it certainly has made its interest known. Here, I do not discuss the many bills that have been introduced in Congress to attack the problem. I do note, however, that a bill introduced by Representative Louise Slaughter (R-New York) (the Genetic Information Nondiscrimination in Health Insurance Act of 1997) would if enacted extend the protections against genetic discrimination in health insurance to virtually every niche that HIPAA overlooked. Another bill, the Genetic Justice Act of 1997 introduced by Senator Tom Daschle (D-South Dakota), focuses on workplace issues. It will be interesting to see whether the congressional focus moves beyond health insurance into broader arenas such as those considered by Senator Domenici and Senator Daschle.

Persistent Public Concerns

Given the paucity of evidence of genetic discrimination and the impressive efforts by state and federal governments to thwart its appearance, I am puzzled that so many consumers, physicians, bioethicists, and others persistently voice fear that the public is unprotected. It may be that the recent positive developments have not had a chance to percolate through society and change opinion. Another possibility is that President Clinton's decision to discuss the threat of genetic discrimination on several occasions—at a commencement address at Morgan State University in May 1997, in an editorial in *Science* in July 1997 (Clinton 1997), and in a speech to a labor union in August 1997—exacerbated public fears of genetic discrimination just when HIPAA should have been (appropriately) reassuring them. In his *Science* editorial, Clinton (1997) wrote:

> Genetic testing has the potential to identify hidden inherited tendencies toward disease and to spur early treatment. But that information could also be used, for example, by insurance companies and others to discriminate against and stigmatize people.

Since he made his remarks a year after HIPAA was enacted, they certainly could lead reasonable people to conclude that he worried about the threat of genetic discrimination in areas other than health insurance.

Some of the most visible proponents of the argument that a threat of genetic discrimination in health insurance persists hold leadership positions in major advocacy groups such as those concerned with breast cancer. They are fully aware of HIPAA, the EEOC guideline, and other positive developments. I have heard them and others argue that the provisions of HIPAA are inadequate to protect against genetic discrimination in health insurance. It is true that HIPAA neither embraces the uninsured nor controls costs, but I fail to see why genetic testing should be a whipping boy for our collective failure to create universal access to health care coverage.

The media, especially over the past two or three years, have discouraged citizens from genetic testing in two ways, one quite accidental, the other quite deliberate. For more than a decade, genetics has been the grist for countless articles about the wonders of scientific advance. Weekly reports of discoveries of the gene for this and the gene for that may have inadvertently contributed to a growing public belief in genetic determinism, although little evidence exists to prove that. I do think that the notion that DNA is a powerful molecule that shapes behavior has, often in humorous ways, permeated popular culture. One now regularly finds references to the "shopping" gene, the "thrifty" gene, and other biological absurdities. Given all the attention the press has paid to genetics, I am not surprised that the public might be easily persuaded and frightened by those who assert that third parties could use genetic information against the public's interests. Hundreds of popular articles warning of genetic discrimination rely at best on flimsy evidence.

One example of that genre appeared in the nation's leading science journal dedicated to a literate public,

Scientific American, in March 1996, under the cover lead, "Gene-Testing Nightmares." In that article Tim Beardsley (1996), the staff writer, asserted:

> Compelling evidence demonstrates that children, whom genetic counselors agree should not be tested for certain traits, in fact are being tested in substantial numbers, to their likely detriment. Conversely, because people at risk for a genetic condition are often turned down for either health insurance, life insurance or employment, patients are now declining genetic testing for themselves or their children, even when it would be medically valuable.

I stop short of saying that Beardsley's statement is irresponsible, but I note that he offered no evidence to support his assertion. I doubt that he could prove his claim. In fact, I think he based it on research in which I was involved that I would argue does not support such a conclusion (Wertz and Reilly 1997). Beardsley's article also featured a photo of one of the leaders of the National Breast Cancer Coalition under an outsize quotation (referring to predictive testing for breast cancer risk): "We will fight any sale of this test before there is consensus on how it should be used." I am troubled that hundreds of thousands of readers of *Scientific American* may have had their first exposure to BRCA1/BRCA2 testing, to me one of the most exciting medical developments of the decade, cast in such a negative light.

Conclusion

I have argued that little evidence exists to support claims that genetic discrimination is a significant problem in our society, that, even so, a sizable fraction of the American public believes that genetic discrimination is a major factor to consider when deciding whether to be tested, that state legislative efforts to control the uses of genetic information by third parties are inadequate (although laws enacted since 1995 do provide individuals with more control over their genetic information than did earlier statutes), and that recent federal initiatives, notably actions taken by the EEOC and the enactment of HIPAA, are quite helpful in protecting against a threat that has yet to materialize.

We are entering an era when genetic testing will become ever more important in medicine. It is quixotic to think that genetic test information can be sequestered from medical records or that it can be absolutely shrouded in secrecy. That could be done successfully with HIV testing only because the test result is so discrete. In the not so distant future, it may well be good medical practice to compile genetic information to assess a broad array of risks. As is the case with many existing biochemical tests, it will be cheaper and may make more clinical sense to amass information through a standard battery. Genetic data will be essential to decisions as broad and diverse as prenatal diagnosis and assisted reproduction, preventive screening, disease prognosis, choice of therapy, and follow-up monitoring. Genetic data will be part of virtually all medical records and will be essential to formulating a strategy to maintain health.

As part of the general medical record, genetic information will be accessible to a large number of viewers, including third-party payers and the administrative staff at health maintenance organizations. How then can we use genetic information to maximize human health and minimize the threat of discrimination and abuse? The best course is not to seek innumerable laws that take on isolated aspects of broad issues. Rather, it is to secure laws that protect the privacy and control the use of all medical information and that dissuade insurers, employers, and others from using information to exclude selected individuals. The Medical Records Confidentiality Act of 1995 is an example of a bill that proposed a reasonable approach to the privacy issue. If one added the category of genetic information to the bill's list of protected entities, a bill like it, shorn of unnecessary bureaucracy, could secure most the goals sought by the more focused genetic discrimination laws.

I am saddened that as we sail forth on what is surely one of the greatest explorations ever undertaken by mankind—the mapping of the human genome—the discovery of information that promises to revolutionize medicine is greeted with suspicion, if not hostility. It may be, however inefficient, good that our society has spent a lot of energy worrying about and responding to the potential threat of genetic discrimination. But enough is enough. If the threat of genetic discrimination has not materialized and is not likely to do so, then we should acknowledge that fact. We should then chart a course to reassure people that fear of denial of access to or loss of health or other insurance should not be among their reasons to forgo a genetic test that can ascertain disease risk and, potentially, save lives.

References

American Council of Life Insurance. *Monitoring Attitudes of the Public.* Washington, D.C.: American Council on Life Insurance, 1995.

Americans with Disabilities Act, 42 U.S.C. §§ 12,101–213 (1994).

Annas, George J., Leonard H. Glantz, and Patricia A. Roche. *The Genetic Privacy Act and Commentary.* Boston: Health Law Department, Boston University School of Public Health, 1995.

Association of British Insurers. *Policy Statement on Life Insurance and Genetics.* February 18, 1997.

Beardsley, Tim. "Vital Data." *Scientific American* 274 (March 1996): 100–105.

Billings, Paul R., Mel A. Kohn, Margaret de Cuevas, Jonathan Beckwith, Joseph S. Alper, and Marvin R. Natowitz. "Discrimination as a Consequence of Genetic Testing." *American Journal of Human Genetics* 50 (March 1992): 476–82.

Clinton, Bill. "Science in the 21st Century." *Science* 276 (June 27, 1997): 1951.

Geller, Lisa N., et al. "Individual, Family, and Societal Dimensions of Genetic Discrimination: A Case Study Analysis." *Science and Engineering Ethics* 2 (1) (1996): 71–88.

Genetic Information Nondiscrimination in Health Insurance Act of 1997, H.R. 306, 105th Cong., 1st Sess. (1997).

Genetic Justice Act of 1997, S. 1045, 105th Congress, lst Sess. (1997).

Genetic Privacy and Nondiscrimination Act of 1997, S. 422, 105th Cong., 1st Sess. (1997).

Health Insurance Portability and Accountability Act of 1996, Pub. L. No. 104-191, 110 Stat. 1936 (1996) (codified as amended in scattered sections of 18, 26, 29, and 42 U.S.C.).

Hietala, Marja, Anu Hakonen, Arja R. Aro, Pirkko Niemalä, Leena Peltonen, and Pertti Aula. "Attitudes toward Genetic Testing among the General Population and Relatives of Patients with a Severe Genetic Disease: A Survey from Finland." *American Journal of Human Genetics* 56 (June 1995): 1493–500.

Hudson, Kathy L., Karen H. Rothenberg, Lori B. Andrews, Mary Jo Ellis Kahn, and Francis S. Collins. "Genetic Discrimination and Health Insurance: An Urgent Need for Reform." *Science* 270 (October 20, 1995): 391–93.

Johnson, J. A. "CRS Issue Brief: Genetic Screening." Congressional Research Service, Library of Congress. Washington, D.C.: Government Printing Office, 1991, pp. 1–15.

Lapham, Virginia E., Chahira Kozma, and Joan O. Weiss. "Genetic Discrimination: Perspectives of Consumers." *Science* 274 (October 25, 1996): 621–24.

Leary, Warren E. "Using Gene Tests to Deny Jobs Is Ruled Illegal." *New York Times,* April 8, 1995, p. 12.

McEwen, Jean E., and Philip R. Reilly. "State Legislative Efforts to Regulate Use and Potential Misuse of Genetic Information." *American Journal of Human Genetics* 51 (September 1992): 637–47.

McEwen, Jean E., Katharine McCarty, and Philip R. Reilly. "A Survey of Medical Directors of Life Insurance Companies Concerning Use of Genetic Information." *American Journal of Human Genetics* 53 (July 1993): 33–45.

———. "A Survey of State Insurance Commissioners Concerning Genetic Testing and Life Insurance." *American Journal of Human Genetics* 51 (October 1992): 785–92.

Medical Records Confidentiality Act of 1995, S. 1360, 104th Congress, lst Sess. (1995).

National Center for Genome Resources. "National Public Opinion Poll on Genetic Testing." Santa Fe: National Center for Genome Resources, 1997.

Reilly, Philip R. "Laws to Regulate the Use of Genetic Information." Pp. 369–91 in *Genetic Secrets: Protecting Privacy and Confidentiality in the Genetic Era,* edited by Mark A. Rothstein. New Haven: Yale University Press, 1998.

Rothenberg, Karen H. "Genetic Information and Health Insurance: State Legislative Approaches." *Journal of Law, Medicine & Ethics* 23 (Winter 1995): 312–19.

Wertz, Dorothy C., and Philip R. Reilly. "Laboratory Policies and Practices for the Genetic Testing of Children: A Survey of the Helix Network." *American Journal of Human Genetics* 61 (November 1997): 1163–68.

Westin, Alan. *Report on National Opinion Surveys Relating to Genetic Testing, Screening, and Uses of Genetic Information, with Implications for a National Survey Focused on Privacy Issues and Safeguards.* ELSI Program (Part 2), U.S. Department of Energy (1997), pp. 1–22.

SIX

Comments on Philip R. Reilly's "Genetic Discrimination"

Ellen Wright Clayton

I very much appreciate the opportunity to respond to Philip Reilly's extremely provocative assessment of genetic discrimination. He makes some points with which I strongly agree, namely that the Health Insurance Portability and Accountability Act (HIPAA) is a big step in the right direction, that most state statutes do little good and may in some cases actually do harm, and that it is hard to say why genetic discrimination is worse than other discrimination permitted within the insurance system and even harder to set up a system in which genetic information is somehow kept from insurers. But I felt like Alice in Wonderland, where black is white and up is down, when I read his assertions that genetic discrimination is not, will not be, and really has not been a problem and his strong suggestion that we should go full steam ahead with genetic testing.

It is true that Paul Billings and his colleagues (1992) have been able to document only a few cases of genetic discrimination in their work. It is not true, however, that their studies were biased in favor of finding cases. In both their studies, respondents had to mail in their stories, a strategy that is almost sure to get a poor response rate, especially when more is asked than simply checking a few boxes. Dr. Reilly points to the lack of response by health care providers to those questionnaires. Their unwillingness to write down their troubles with their patients' insurers is hardly surprising to me; they already spend hours every week advocating with insurers for things their patients need. Everyone I know who cares for patients is worried that knowing about health risks is bad for access to health care.

Another point to be made here is that predictive genetic testing of asymptomatic individuals is still in its infancy. Very few people at present actually know their risk of developing a late-onset disorder. It may not be surprising, therefore, that it is hard to find people who have been denied access to employment or insurance as a result of their genetic predispositions. But insurers, fearful of adverse selection, have repeatedly stated that they want to know whatever those they insured know. And they will use such information to adjust rates, to deny coverage, or to exclude certain conditions, all of which is legal, as Dr. Reilly pointed out, but not all of which is just or socially desirable.

While HIPAA is a big step in the right direction, it does not come nearly so close to solving all the problems as Dr. Reilly suggests. People who seek coverage in the individual market are not the only people who still have problems, and it is not true that most of those people can pay whatever it takes. The fact is that major holes remain in the coverage provided by the act. Millions of people work for small employers, who, in turn, are subject to medical underwriting. If an insurer finds that one employee in a small group is a high risk, the insurer is still permitted to raise the rates charged to the employer and to exclude some interventions from coverage. The employer in that case can and increasingly does choose simply to stop offering health insurance to its workers. Another hole is that employees who transfer from one type of group insurance to another cannot avoid the preexisting condition exclusions and some other adverse conditions if they have a break in coverage of more than sixty-two days. To be sure, many employees change from one job to another in a relatively seamless manner, but millions of workers experience prolonged periods in which they are between jobs, some of which exceed the sixty-two days. HIPAA is a good thing, but tremendous potential still exists to use information in ways that act to the detriment of the insured. Fixes may be on the horizon, or even pending in Congress, but they are not law yet.

I would like to make one final point about discrimination, this time in the context of employment. Dr. Reilly has already pointed out some of the potential weaknesses in the application of the Americans with Disabilities Act to genetic predispositions. But even if the ADA does appropriately apply to individuals who are predisposed to get sick, it is quite difficult for plaintiff-employees to prevail in claims that they were subjected to unwarranted discrimination in the workplace.

The other major point I want to make is this: I share Dr. Reilly's excitement about some of the new developments in human genomics. Finding mutations that predispose individuals to develop cancer or other serious disorders creates the opportunity to learn more about the pathogenesis of those diseases and, one hopes, to develop ways to prevent or treat them. To investigators, the idea, for example, that many types of cancer occur when mutations occur in genes that normally clean up naturally occurring mistakes suggests a number of possible therapies.

But for most of those diseases, effective interventions are not yet available. What is more, the prognostic implications of having certain mutations are not well defined either. We have good reason to think, for example, that mutations in the same gene may have different effects in different families. To pick just one example, in high-risk families, 85 percent of women who have mutations in BRCA1 will develop breast cancer during their lifetimes. By contrast, Ashkenazi Jewish women who have the 185delAG mutation in that gene appear to have a 60 percent risk of developing breast cancer, and women in that group who have the mutation but no family history of breast cancer may have an even lower risk.

When so much uncertainty exists about what having a mutation means and what can be done to avert the risk of becoming ill, one cannot simply assume that genetic testing is a good thing. To be sure, individuals who find out that they do not carry the "family mutation" are likely to be relieved, issues of survivor guilt and personal redefinition notwithstanding. But the effect can be terrible for those who do not get such good news and who must then live with all the uncertainty. One could argue that we should go ahead with testing, assuming that so long as people know what they are getting into, they should be free to do as they like.

I, however, agree with the recommendations of the Task Force on Genetic Testing that genetic testing not be made widely available until the results have been demonstrated to be useful for patients. I have several reasons for saying that. First, we have learned from the experience with Huntington's disease that when people at risk really understood what the tests could and could not do, the overwhelming majority chose not to be tested and to continue to live with the question of whether they were truly at risk. A fifty-fifty chance that they had dodged the bullet was better than knowing that they had not. Second, despite my work on developing new ways to educate patients about genetic tests that do not rely on the ability of the primary-care practitioner, I remain skeptical about our ability truly to help people decide whether to be tested, particularly when so many practitioners understand so little about genetics. Finally, the history of medicine is replete with instances in which new tests and therapies were introduced with great enthusiasm only to fall into disfavor as experience demonstrated that their promise was overblown. I should make clear here that my wish that we learn from history and require greater showing of efficacy before new interventions are introduced into medicine is not limited to genetic technologies but to all medicine. Finally, genetic testing for disorders offers great promise, but we would do well to define the proper role of genetic testing in the medical armamentarium and to ensure that people can actually benefit from the information they gain from testing before offering those tests in routine practice. That time may well come, but it is not here yet.

References

Billings, Paul R., Mel A. Kohn, Margaret de Cuevas, Jonathan Beckwith, Joseph S. Alper, and Marvin R. Natowitz. "Discrimination as a Consequence of Genetic Testing." *American Journal of Human Genetics* 50 (March 1992): 476–82.

Reilly, Philip R. "Genetic Discrimination." In *Genetic Testing and the Use of Information,* edited by Clarisa Long. Washington, D.C.: AEI Press, 1999.

Index

INDEX

The American Enterprise Institute for Public Policy Research

Founded in 1943, AEI is a nonpartisan, nonprofit research and educational organization based in Washington, D. C. The Institute sponsors research, conducts seminars and conferences, and publishes books and periodicals.

AEI's research is carried out under three major programs: Economic Policy Studies; Foreign Policy and Defense Studies; and Social and Political Studies. The resident scholars and fellows listed in these pages are part of a network that also includes ninety adjunct scholars at leading universities throughout the United States and in several foreign countries.

The views expressed in AEI publications are those of the authors and do not necessarily reflect the views of the staff, advisory panels, officers, or trustees.

www.ingramcontent.com/pod-product-compliance
Lightning Source LLC
Jackson TN
JSHW011938131224
75386JS00041B/1450